C. R. Boyd

Resources of South-West Virginia

Showing the Mineral Deposits of Iron, Coal, Zinc, Copper and Lead

C. R. Boyd

Resources of South-West Virginia

Showing the Mineral Deposits of Iron, Coal, Zinc, Copper and Lead

ISBN/EAN: 9783337181086

Printed in Europe, USA, Canada, Australia, Japan

Cover: Foto ©berggeist007 / pixelio.de

More available books at **www.hansebooks.com**

HALE'S FALLS, GRAYSON CO., VA.

(P. 308.)

RESOURCES

OF

SOUTH-WEST VIRGINIA

SHOWING THE

MINERAL DEPOSITS OF IRON, COAL, ZINC,
COPPER AND LEAD.

ALSO,

THE STAPLES OF THE VARIOUS COUNTIES, METHODS OF TRANS-
PORTATION, ACCESS, Etc.

ILLUSTRATED BY NUMEROUS PLATES AND LARGE COLORED MAP
REPRESENTING THE GEOGRAPHY, GEOLOGY AND
TOPOGRAPHY OF THE COUNTRY.

BY

C. R. BOYD, E. M.,

MEMBER OF AM. SOC. OF CIVIL ENGINEERS, AND OF THE INSTITUTE OF MINING ENGINEERS.

NEW YORK:
JOHN WILEY & SONS,
15 ASTOR PLACE.
1881

NEW YORK : J. J. LITTLE & CO., PRINTERS,
11 TO 29 ASTOR PLACE.

CONTENTS.

WYTHE COUNTY.

CONTENTS.   vii

GILES COUNTY.

RESOURCES

OF

SOUTHWEST VIRGINIA.

It is hoped by the writer that the matter contained in this book will be a sufficient apology for its appearance. Not that he is, by this means, trying to forestall criticism upon the manner in which the subjects are handled. No doubt a much more beautiful structure could be built of the same materials by more skillful hands, but *now* is the time when the general public desire to be informed about these most lavish and bountiful resources of this section of Virginia ; and if the more competent heads and hands will not undertake a work which, if even tolerably well done, would be alike gratefully received by a progressive public and a needy State, why some one must do it who may barely have nothing more than a love of his country to urge him to the task.

The great and crying need of Virginia now is not so much the continuance of a suicidal strife over an issue already determined by the highest law and usage, but that we should forget all animosities, and labor together to build up and largely enhance the financial power of such communities in the State as are capable of further development. This can be done by showing up our resources in a proper manner ; thus bringing in many men of capital to willingly help us not only bear our burdens, but create new facilities for making money, in the erection of furnaces, etc., and in the build-

1

ing of new lines of railway. It is not any attempt at prophecy to say that the financial power of Southwestern Virginia and of the James River Valley in ten years will be increased thirty-three per cent. If we of this section can contribute anything to hasten the good time, why then let us do it. The facts truthfully stated, a picture or two here and there of the fine scenery, a measurement now and then of the vast mineral veins and lodes with which Nature has so lavishly blessed this country, and an earnest endeavor to bury self a little while, and resurrect the country, will do the work for us.

This work, then, upon the resources of a section so rich naturally, needs no introduction, other than an apology for the great imperfections which must mark an attempt to treat so important a subject in a manner it deserves.

Some of the dear friends of the writer desire that he will use the occasion to express his views upon the great question which is said to have such a bearing upon the speedy development of the massive resources described in this volume: namely, that of the tariff; but he begs to defer such a discussion to a more suitable occasion. Likewise it has been contended that the great purity and excellence of the ores of iron described, and cheap and abundant fuel, mark the section as the one to which the attention of government officials should be timely directed, in anticipation of the great efforts likely to be made to bring up the naval armament of the country to a condition of high efficiency; but, though it is to be regretted that such friends will be disappointed here, the writer begs to submit the impropriety of loading the book with arguments and opinions that would obscure a clear view of the subjects treated, besides creating prejudices in advance against the territory described, that would remand both the noble area thus treated and the writer, to a lasting want of appreciation.

MONTGOMERY COUNTY.

It may be of interest, before going into a description of Montgomery County as it now looks on the map, to revert to its earlier history, speaking of its origin.

Augusta County, which once contained all that there is now of Southwest Virginia, was created in 1738 from Orange County, a scope of country which comprised not only the territory that this book treats of in Virginia, but the whole of the State of Kentucky. Bottetourt County came next, being taken out of "West Augusta;" and then, in 1772, Fincastle County was formed, covering our ground. A short time after, in 1776, the county of Fincastle was extinguished by the formation of Washington, *Montgomery*, and Kentucky counties, the last comprising the present State of Kentucky.

It is needless to remind many of the readers of this book, that General Washington, when he spoke of the possibility of having to retreat to the mountains of "West Augusta," alluded to the chains of mountains which pass through the heart of Southwestern Virginia.

Montgomery, shorn of much of its original territory by the formation of several new counties on every side, may be said, now, to lie between the crest of the Gap or Walker's Mountain on the north, and Laurel Ridge and Bent Mountain on the south; bounded west, practically, by New and Little Rivers, and east by no very marked geological data, running parallel with the line between it and Roanoke County.

GEOLOGY.

The geology of the county is comprised between the Huronian rocks, lying on the south, and the proto-carboniferous measures toward the north side, inclusive.

The classification of the rocks and the divisions between different epochs are more or less easily defined, with the ex-

4 MONTGOMERY COUNTY.—GEOLOGY.

ception of some of the limestones in the valleys. These great masses of limestones, which are, very frequently, indiscriminately assigned to the Trenton sub-epoch, are almost totally without fossils of any kind ; and, but for their known relation to very correctly determined data, might as well be assigned to the Eozoic as any other period, so far as fossils are concerned. They lie conformably above the rocks showing that well-known fossil the scolithus, and do really belong to the Calciferous sub-epoch, which throughout this part of Virginia is very thick ; and in it mainly repose the thickest and most valuable deposits of brown iron, zinc, and lead ores, of which this volume is likely to speak: though it is by no means to be understood that any invidious distinction is being attempted, in advance, against the fine brown ores in Numbers VII. and VIII. to be found in the Gap Mountain. It is not here assumed that there are no Trenton limestones in this section of Virginia ! The remarks above apply to the main body of the limestones in this southwestern extension of the Great Valley of Virginia. Underlying the north base of some of the larger mountains, the Trenton is very well represented.

The cross section on the opposite page, twenty-two miles in length, which has been taken across the county from south to north, may lead to a much clearer comprehension of its geology. It will be seen from an examination of it that the Huronian strata, as well as the neighboring rocks of the Potsdam and lower part of the Calciferous, have been slightly overturned toward the north, or partially reversed in their true order. Next to these, going north, are the great limestones of the Calciferous, carrying us over the great iron, lead, and zinc beds. Proceeding northward over a repetition of the broken Calciferous, crossing the line of the Atlantic, Mississippi and Ohio Railroad, a great fault is encountered, which repeats itself several times in Southwestern Virginia.

Gap on Walkers Mountain

Poverty Valley

Brush Mountain

Toms Creek

Valley of Blacksburg

Stroubles Creek

Prices Mountain

Atlantic, Miss. & Ohio R.R.

Great Valley of Virginia

Pilot Mountain

Brush Creek

Laurel Ridge

A Section through Montgomery Co., Va.

Scale of Miles

N 55° W.

TRENTON
ONEIDA
CLINTON
ORISKANY & C. (IRON)
MARCELLUS & HAMILTON
CHEMUNG
CATSKILL
COLEY
PROTO CARBONIFEROUS

UNDETERMINED, APPARENTLY McLUREA LIMESTONE

TRENTON
HUDSON
ONEIDA
HAMILTON
CHEMUNG
PROTO CARBONIFEROUS

Surface showing Talc-through Limestone, rather "horizontal" in the centre of the basin but generally at all angles.

GREAT IRON BELT

CALCIFEROUS POTSDAM

LEAD ZINC
GOSSAN
GOLD THIN
FELSPALMIG
QUARTZOSE GOLD
MICACEOUS
GOLD
GOLD

That is, we have the rocks of the Calciferous brought up into
contact with a downthrow of the proto-carboniferous, or lower
coal rocks. Passing this fault, the measures holding the
Price's Mountain coal are soon encountered, lying in the shape
of an anticlinal ; then the limestones of the beautiful valley,
watered by Tom's Creek and Strouble's Creek principally ;
then the Brush Mountain coal measures. After which the
Olean conglomerate, underlaid by a great width of slates,
shales, and sandstones, through the entire upper petroleum
rocks, upturned and visible. Then reaching the base of Gap
Mountain, the Marcellus black slates and shales, underlaid by occasional beds of the Corniferous, the Oriskany, the
lower Helderberg, the Clinton, and the Oneida last. These
last mentioned of the Upper Silurian, outcropping almost invariably on the southern slope of the mountain, dip at angles
varying between 60° and 30° southwardly.

BRUSH CREEK.—GOLD-BEARING ROCKS.

The south side of the county is marked, as we have
seen, by Huronian strata. These rocks, mainly micaceous,
felspathic, and chloritic, banded here and there with heavy
dikes of quartz, trend N. E. and S. W. the length of the
county on that side. The decomposition of these rocks
through many ages has not only left a good soil along the
valley of Brush Creek, but now the people of that region are
somewhat excited over the gold which is being found along
the streams. Whether any distinct stratum exists with well-
marked veins of the precious metal has not yet been so fully
determined as desired. The washings so far show the gravel,
etc., to yield about 33 pennyweights to the hand per day,
with rude sluice boxes. These same strata cross the South
Fork of Roanoke River, about the line between Montgomery and Floyd Counties, and ought there, in the deeper

gorges, to yield more heavily than anywhere else. To speak more to the point, by looking at the general map accompanying, Pilot House will be found on the southern side of the county of Montgomery on Brush Creek. Three miles east of Pilot House, more gold has been washed from the gravel and other detritus than at any other point. The gold has been found in surface washings over an extent of country six to eight miles wide, chiefly in quartz dikes in mica slates alternating with granite which is occasionally porphyritic. The dip of the rocks is north here, but on the south side of Laurel Ridge the dip is the other way. Gold has been found in very handsome quantities on Laurel Creek.

The rock ledges near Pilot House are granitic next to Pilot Mountain; near the foot of this mountain, next to Brush Creek, hydro-mica slates come in, showing here and there ledges of quartz, one of which, from six to twelve feet thick, holds the most of the much-talked-of gold of Brush Creek. Then south of this are the hydro-mica, chloritic, and schistose bands, which are again repeated in Laurel Ridge, giving the gold again. Then again on Laurel Creek in Floyd County.

A large vein of *pyrites*, containing *copper*, exists undeveloped in this valley, as well as *lead ore*. Occasional fragments of magnetite are also found. Next to the north is *Pilot Mountain*, the eastern prolongation of Iron Mountain, occupying the line of junction between the Huronian and Cambrian rocks, that is between the metamorphic and more distinctly fossiliferous, such as the Potsdam.

This mountain is cut off just east of the prominent and well-known peak called Fisher's View, by the south fork of Roanoke River, and rises again farther east in a series of high and steep spurs known as Poor and Bent Mountains. In the northern escarpment of this range is the division be-

tween the Potsdam and calciferous sub-epochs, marked by a
vein of

BROWN IRON ORE,

trend N. E., from 15 to 30, and sometimes 100 feet thick—
a decomposition of pyrites. Here and there, owing to end
compressions and side flexures, this vein assumes a much
greater thickness. For instance, a flexure in the Big Bend
of Little River, near Elliott's, caused the vein to have a
measure through a hill of about 300 feet. This mass will
yield about 400,000 tons of ore in 375 feet length, above
water level.

To attempt to estimate the quantity this vein the whole
length of twenty miles in the county would yield, would be
absurd. Above water level it would give a breast of several
hundred feet elevation, approaching the perpendicular in
attitude. Near Tice's Mill, South Fork of Roanoke River,
the comparatively deep gorge takes us down below the decomposed zone, and you find the vein of pyrites nearly 100
feet thick.

In many places this great vein is charged with manganese
to such an extent as to render the iron made from it very
hard. Here and there it is phosphoretic; but as an ore for
general purposes, to be mixed with other ores, it is very
good. Now and then it assumes the form of pipe ore in some
of the hills, becoming stalactitic and mamillary in form,
and these ores usually are among the best in the great Iron
Belt.

In other sections there are generally two iron veins of
lesser thickness between the one just described and the
measures holding the lead and zinc, but in this county they
are not so distinct. *The lead and zinc* are not more than 900
yards, across alternate strata of red and gray shales and
limestones, northwardly from the great iron vein in Mont-

gomery County. Beginning at Calfee's in Little River District, and continuing northeastwardly through the mine opened by Col. Langhorne of Shawsville, are here and there interesting exposures of the vein. Proceeding on eastwardly toward Big Spring, the exact locality of the vein is lost under the great amount of debris in the foot hills of the higher mountains. At Calfee's and at Colonel Langhorne's the exposures made show the vein to be nearly on its edge, between walls of limestone from 9 to 12 feet thick, and yielding very handsome specimens of lead sulphuret and zinc blende and carbonate.

After leaving this measure, going northwardly, you pass across several hundred yards of Calciferous red slates and shales and magnesian rocks ; the reds now and then giving fine crystals of *specular red iron ore.* From the decomposition of these rocks must result a great part of the irregular deposits found in different parts of the limestone valley farther down and somewhat north.

The limestones toward the middle N. E. and S. W. belt of Montgomery are so well known as the limestones of the Great Valley of Virginia as not to require any great amount of description. In them occur beds and deposits of brown iron ores. In one series about three miles south of the Atlantic, Mississippi and Ohio Railroad, which runs east and west through the middle of the county, there are surface indications of *magnetic ore.* It has not yet been developed. Just south of this line is a series of easily decomposed limestones, running the length of the county, from which has resulted a great deal of crystalline lime carbonate, sometimes having the appearance and texture of Mexican onyx.

North of the line of the A. M. and O. R. R.* the fault is soon encountered which separates the great valley limestones from the coal rocks of the western end of the county

* Since the book has been in press, this road has passed into the hands of the " Norfolk and Western Railroad."

and from the Devonian limestones, slates, shales, etc., of the eastern end, which correspond in position to the coal rocks.

In this line of Devonian Rocks occur the celebrated springs of Montgomery, with the exception of Alleghany Springs, which are situated at the base of the mountains, holding the earlier formations, on the south side of the above-named railroad. Of all these mention will be made in detail presently.

The coal rocks, which begin to show in the southern spurs of Price's Mountain, three miles N. W. of Christiansburg, on the lands of Kyle and others, belong to the earlier sub-carboniferous, or proto-carboniferous ; and, for that side of Price's Mountain, while here and there sufficiently regular to yield a good return for mining, it is not until you reach the northern escarpment of this mountain and the southern side of Brush Mountain, inclosing the valley of Blacksburg, that you reach the better areas.

THE VALLEY OF BLACKSBURG.

Among the mere exhibitions of beautiful landscape scenery, presented so frequently to the eye in traveling over South-west Virginia, perhaps no scope of the whole extent will so forcibly remind the beholder of a fine English landscape as the valley of Blacksburg : swelling and undulating grassy meadows and grain fields, relieved by occasional extensive areas of woodland, in which the oak almost invariably out-numbers all other trees put together ; numerous streams, flowing from springs in the limestone, sometimes so bold as to afford power sufficient to run a grist-mill within a few hundred yards of the source ; mountain chains, on the north and south, with its western limits defined by a broad and beautiful river.

Section through Blacksburg Valley, Montgomery Co., Va.

Two veins of good Coal, having about the same character in both mountains, by measurements taken underground.

The small vein, 30 ft. under the large vein, is 3 ft. thick and of good Coal.

BLACKSBURG COLLEGE.

In this valley, near its eastern side, are the town of Blacksburg, and that State institution of learning known as " The
Agricultural and Mechanical College." The reader, it is
hoped, will excuse us for a moment, if we delay the discussion of the geology of this remarkable section, in order to
say a word concerning Blacksburg College. This college,
erected by the State at a time when her financial embarrassments were such as to render it impossible for her to put
enough money into it to either complete it in a manner to
meet the objects for which it was organized, or to maintain
it properly when so built, has practically proven the grave
of an estimable faculty. Without wasting more time upon
the discussion of disagreeable matters, it is a question
whether or not it would be advisable for the State to consider the propriety of turning this college into an asylum of
some sort—an institution which she has also been trying to
secure—to be located in that part of the State; and then,
taking advantage of the fact that Emory and Henry College
owes the State a large sum of money, to buy out that institution with this debt and the proceeds of the sale of the State
farm at Blacksburg; thus not only securing the admirable
grounds and buildings of Emory and Henry College, but
rendering it possible to make up such a faculty out of the
two schools combined as will make an entire success, as a
whole, out of the two, whose hope of a successful future
apart is entirely chimerical.

If such a solution is possible, Blacksburg and vicinity in
the matter of dollars and cents must be gainers by the
change. The overcrowded condition of the asylums in different parts of the State would render it necessary to send
more persons there to consume the produce of the surrounding country than they now have to supply at the college, or

are likely to have for many years to come. On the other hand, Emory being now brought down in the number of its students by the erection of Vanderbilt University and other colleges west, after becoming a State institution, under this plan, would have a great accession in numbers in the State students, as well as many others attracted by the greater facilities for a liberal and cheap education it would then present, compared with its present status.

To those interested, who will follow this line of reasoning out in all its bearings, it must be evident that on all sides there would be great gain to State, to faculties, and the public. . . . To resume the geological features again, this valley of Blacksburg presents quite an interesting problem to scientific men in her *coal field*. Do the coal measures which outcrop on either side of this valley form a lenticular basin underlying the valley, or is it broken in the middle by an intrusion of silurian limestones from below? Let us here insert an accurate cross section, taken directly across the valley, to assist in the discussion, a distance of just 3.34 miles from outcrop on the south side to the corresponding strata on the north side.

It has been assumed by Professor Lesley, of Pennsylvania Geological Survey, that the limestone in the valley is lower silurian; consequently, that there has been such a disturbance of the earth's crust as to force a great fragment of this limestone into a rent made in the coal formation. This may be true; but a very close examination has reduced the probable lateral thickness of this limestone of the lower rocks to only one mile or less. Beginning at the outcrop of the coal on the south or Price's Mountain side, 6,000 feet were measured before the coal rocks, which could be identified with some certainty, were passed; and then on the north side, within about 6,000 feet of the outcrop in Brush Mountain, a repetition of the same rocks, dipping southwardly, was encountered, making 12,000 feet out of 17,625 feet, which was the whole length from one coal outcrop to the other of the basin, toward the center of which the coal dips from each side.

The general absence of abundant fossil remains in the limestones so far as you can see them would naturally lead to the belief that they belong very low down in the order of true position ; but they may belong to that class of rocks which are formed as a precipitate sometimes in waters rather too deep for animal life of any kind. In that case they may as well be referred to one age as another. After a tolerably diligent search there were a few fossil remains found, but too indistinct to be regarded as positive evidence. Remains of what was without doubt either a McLurea, a Goniatites or Clymenia were found as you approach the northern side of the valley, while on the southern side some of the limestone was brecciated with distinct crystals of gypsum, an almost conclusive proof of their being Carboniferous ; with a considerable measure of these limestones not only lying conformably upon the well-recognized red and gray shales of the coal measures, but differing very materially in structure from those near the center of the valley. They were crypto-crystalline over much of the ground where much exposed, as though they had been largely composed of fragments of gypsum which had decomposed and passed away in solution, leaving the mass full of cells. But it is not proposed to render any decision here, or to prolong the discussion to the exclusion of other matter which may be more agreeable to the reader ; but it is a question submitted to men of science as to whether this is a true basin or not. If it is broken how is the intrusion of the lower rocks into so small an aperture, comparatively, without bringing along with them much of the intervening formations, to be accounted for ? Industrially, these coals are valuable. Four veins, the larger having fifty-four inches of good coal, in a vein nearly nine feet thick, the smaller two and one-half feet of good coal in a vein three feet thick ; the smaller being much more bituminous than the larger, which is an anthracite coal.

These veins lie very much as represented in the cross section, coming in again on the south side of the anticlinal formed by Price's Mountain. In Brush Mountain the most satisfactory measurements were taken at the Kinser Bank and at the Faulkner Hollow. In Price's Mountain, Bruce's Mine gave the most accessible openings. From these measurements the identity of the veins was established, and as they varied so little at different points, it may safely be assumed that the measures given are the proper ones upon which to base calculations as to quantity.

These two veins, the only ones of any consequence, are much broken up and disturbed from New River, going east, some distance beyond Poverty Gap, leaving ten miles length, however, of regular measures, beginning as you approach Price's Mill, and becoming broken again in the vicinity of Blacksburg : ten miles on either side—both Brush Mountain and Price's Mountain. These veins will average at their outcrops 180 feet above water level in Tom's and Strouble's creeks, judging from careful barometric reading— average dip 25 —giving 426 feet on the length of the incline. Specific gravity being 1.65, the quantity in the larger vein above water level in these creeks will be found to be over 4,600,000 tons on each side of the valley ; deducting one-third for space taken up by ravines and loss from handling, there would be 3,000,000 of tons for shipment. In the three feet vein, as remarked above, there will be two and one-half feet of coal of a more bituminous character ; this will yield for each side in the same distance over 2,500,000 tons above water in the same creeks, or over 1,600,000 tons for shipment.

There would be yet remaining the coal below water level, which would run on the inclination, or dip, of the veins at least 5,000 feet on each side of the valley, and would yield a tonnage more than ten times as great as that enumerated above. Other parts of these veins will yield many hundreds of thousands of tons, both as you approach New River, and as you go toward the valley of the North Fork of the Roanoke ; but those sections, as well as the south side of Price's Mountain, will never be regarded as reliable as that from which the above calculations were made.

The analysis of these coals may not be uninteresting :

Coal from the large vein analyzed by J. M. Safford, State Geologist of Tennessee, in 1876, gave the following results :

Moisture	0.89
Volatile matter	7.82
Fixed carbon	67.29
Ash of grayish color	24.00
	100.00

Mr. Safford declares it an anthracite.

The coal of the smaller vein, analyzed in 1876, in the laboratory of the University of Pennsylvania by George A. Koenig, Ph.D., professor of metallurgy and applied chemistry, gave the following results:

Moisture at 140° C.	0.20
Gas given out by heating to a yellow-red heat one hour	27.44
Fixed carbon.	57.92
Ashes (white)	14.34
Sulphur	0.10
	100.00

This chemist places this coal in the dry semi-bituminous class. The *whole quantity* of the coal removed from these veins does not exceed 50,000 *tons* to date.

The average quantity per year is about 1,700 tons.

THE MILLSTONE GRIT.

The millstone grit, nowhere more than 245 feet thick, is well exposed in Brush Mountain, near Price's Mill; several ledges of it there yield a stone from which millstones have been made for many years. These millrocks have even been used with satisfaction in the large mills at Richmond ; and would, no doubt, form the basis of a large business, if there was a railroad passing by. Between this millstone grit and the coal there is a ledge of conglomerate, the pebbles of which yield a low percentage of silver upon analysis.

POVERTY VALLEY AND GAP MOUNTAIN.

Passing then into the valley next Gap Mountain, nothing of note is encountered until you reach the limestone of the Lower Helderberg group, which will be valuable for furnace use. It sometimes contains small quantities of blue lead ore. Close to this going north, as you ascend the Gap Mountain, is the remarkable ledge of brown weathered-look-ing sandstone, which all through this section is an iron and sometimes a manganese-bearing rock. Frequently the sand-stone gives way entirely to a limonite highly prized by iron men, making the Oriskany an exceedingly valuable member of the series. In this particular the Oriskany has great supe-riority over the Clinton in this vicinity. The Clinton or Dyestone ore here does not seem to have the character possessed by the same measures in Poor Valley Ridge, Lee County.

Now let us revert to the mineral springs, after stating in

2

brief that the northeastern quarter or division of the county is composed mainly of Devonian rocks, on the waters of the North Fork of Roanoke River, inclosing Lewis Mountain; with the exception of the north bend of the above stream, which is of lower limestone lying next to the proto-carboniferous rocks in that extension of Brush Mountain.

MINERAL SPRINGS.

Alleghany Springs are very pleasantly situated on the south bank of the Roanoke River, three and one-half miles south of Alleghany Depot, a station on the Atlantic, Mississippi and Ohio Railroad. These springs may justly be ranked among the most excellent and desirable summer resorts in Virginia. The waters are highly medicinal in their character, judging from the following thorough analysis:

DR. GENTH'S ANALYSIS.

"According to my analysis, the composition of the water from the Alleghany Springs is as follows:

"ONE GALLON, 70,000 GRAINS, CONTAINS"

Sulphate of Magnesia	50.884200 grains		Nitrate of Magnesia	3.219562	grains
do.	Lime	115.300122 "	do. Ammonia	0.779432	"
do.	Soda	1.717529 "	Phosphate of Alumina	0.025549	"
do.	Potassa	3.088484 "	Silicate of Alumina	0.26789	"
Carbonate of Copper	0.000239 "		Fluoride of Calcium	0.025888	"
do.	Lead	0.000569 "	Chloride of Sodium	0.273656	"
do.	Zinc	0.004713 "	Silicic Acid	0.882782	"
do.	Iron	0.157049 "	Cronic Acid	0.000921	"
do.	Manganese	0.030017 "	Arsenic Acid	0.000192	"
do.	Lime	3.613209 "	Other Organic Matter	1.989121	"
do.	Magnesia	0.302802 "	Carbonate of Cobalt..⎫ Traces.		
do.	Strontia	0.003536 "	Teroxide of Antimony ⎰		
do.	Baryta	0.022304 "			
do.	Lithia	0.001679 "		181.073321	grains

Solid ingredients by direct evaporation gave	181.07380	"
Half combined carbonic acid	1.985561	"
Free carbonic acid	5.155765	"
Hydro-sulphuric acid	0.001539	".

Total amount of ingredients	190.111912	"

"With regard to the medicinal qualities of this mineral water, I beg leave to copy a few passages from a letter of CH. COCKE, M.D., a gentleman who has made for a number of years a series of observations of its healing qualities, and has furnished me with information on this subject. He says: 'I have spent a portion of six seasons at the Alleghany Springs —the two first for the cure of an inveterate case of dyspepsia of twenty-five years' standing—the stomach being the chief seat of the disease, with many sympathetic affections, such as nervous headache, palpitation of the heart, etc. *The cure was perfect.*'

*   *   *   *   *   *   *

"'In short, I have never met with a case of derangement of the digestive organs (where the waters were properly used with the necessary perseverance) that was not cured or greatly relieved, except where the patient was far gone in consumption—a complication of diseases by no means rare.'

*   *   *   *   *   *   *

"The numerous ingredients found in the water of the Alleghany Springs, in small quantities, in connection with the more active salts of lime and magnesia, must certainly produce decided effects, and in combination with the delightful climate, fresh air, and exercise, cannot help but have a highly beneficial influence in many affections of invalids.

"(Signed) F. A. GENTH."

A stay at these springs has often proved highly beneficial to those who have become debilitated from any cause. A most healthy desire for food is one of the marked effects, and with it that most grateful feeling produced by a restored power of digestion which all mere appetizers cannot claim.

This condition, however, may be owing to the pure and wonderfully bracing atmosphere of the place, seeming ever to

be renewed and freshened by contact with the lofty mountains stretching away for miles in the background.

As to its natural beauty! It is picturesque and romantic
in the highest degree! No lover of the beautiful in nature
could fail to be deeply thrilled with the almost perfect harmony in every feature of the noble landscape, presented by
the grounds and buildings of the springs, in their setting of
wooded hills and mountains—the mountains rearing their
towering summits, plumed here and rock-covered there, far
into the unlimited space above, as if wooing the soft embrace
of the fleeting clouds. In its front flows the never-failing
South Fork of the Roanoke, washing the foot of the lawn,
fresh from more than a thousand crystal springs, dancing to
its own sweet music, as it rapidly descends over boulder and
pebble. It may be as well that we have no sketch of these
springs, for no pencil could portray on paper the charming
picture.

Near to these springs is the fine scenery afforded by the
mountain streams in their rapid descent through the deep
gorges—Styles' Falls of fifty feet, Puncheon Run Falls of
three hundred and fifty feet at the steepest, with much of it
a cascade coming down a thousand feet. Add to these, two
falls on this fork of Roanoke River, which will have to be described in the scenery of Floyd County, the Beatrice and
Prince Imperial, and you have a group of attractions very
rare to occur in so short a distance of each other. It is seldom that two streams unite in the manner presented by the
South Fork coming over Beatrice Falls eighty feet, forming a
basin at its foot, into which a smaller stream pours its waters
two hundred feet over the Prince Imperial.

Space and our feeble powers alike forbid us to attempt a
description of all this scenery. Nature will not fail to impress any who may visit these places with their exceeding
beauty more than their awful grandeur. To see them is to

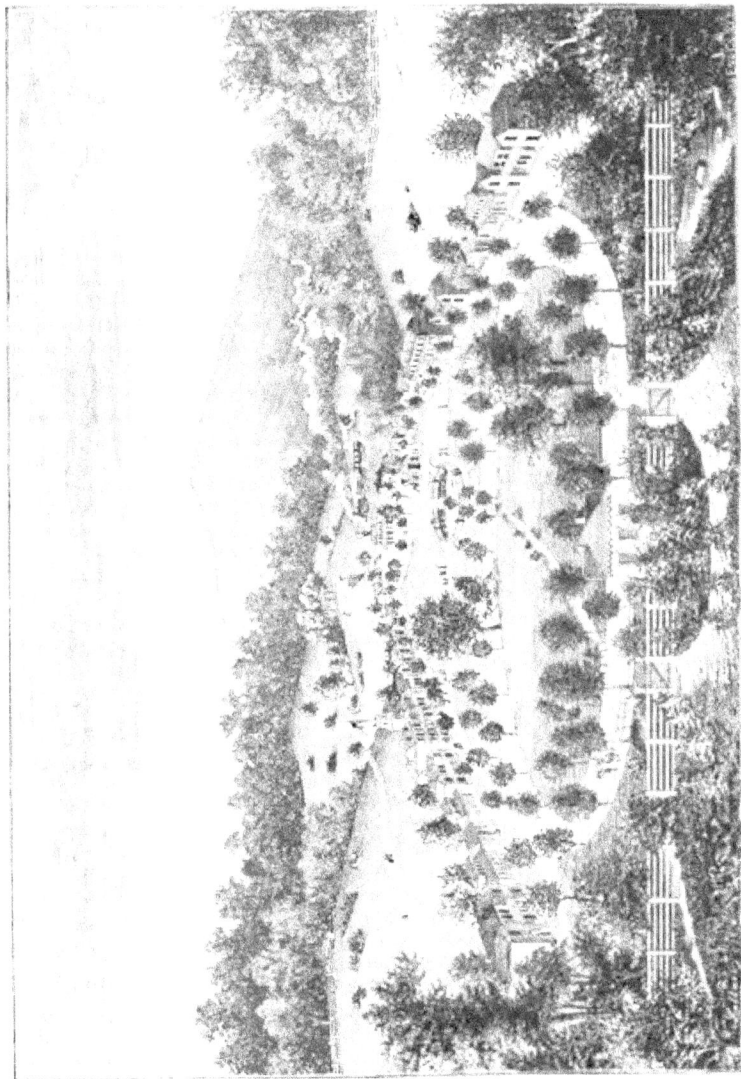

MONTGOMERY WHITE SULPHUR SPRINGS, VA.

(P. 31.)

read what nature has written in a language no pen can describe nor pencil portray.

THE MONTGOMERY WHITE SULPHUR SPRINGS.

These springs derive their waters from Devonian limestones, just about the line of the great fault. The grounds
and buildings are situated in a very attractive spot, about
one and three-quarter miles north of the Big Tunnel, a station
on the Atlantic, Mississippi and Ohio Railroad, with which
they are connected by a narrow-gauge railroad. All that can
be said of pure air, mountain scenery, and excellent diet may
very well be applied to this locality. The waters are of several kinds—three bold sulphur springs, one chalybeate, and
a freestone spring, besides limestone water in abundance.
The handsome pavilions over the marble reservoirs, in which
most of these waters are caught, are not the least attractive
feature of these springs. It is claimed that the hotels and
cottages here will accommodate one thousand guests, and
that they are well supplied with that great luxury absent at
many places—bath-rooms.

In this vicinity, on a stream running into the North Fork
of Roanoke River, are the *Dudley Falls*, a beautiful cascade ninety feet high, where the limestone water, agitated in
its descent, has deposited in so many years great masses of
tufa. On the North Fork of Roanoke River, easily accessible
to these springs, are many wild and romantic dells. In all
these places in summer, or when the trees are loaded with
the frosts of winter, there is a distinct charm that impresses
every lover of nature.

THE YELLOW SULPHUR SPRINGS.

The fine mineral spring to which this place owes its reputation is situated also in the general line of the great fault

before alluded to, and may derive its waters from the materials composing the earlier as well as the Devonian limestones.

The site for its grounds and buildings was well selected in a wooded glen, about three and a half miles northwardly from Christiansburg Depot, on the Atlantic, Mississippi and Ohio Railroad, at an elevation of about two thousand feet above the sea. It is in full view of fine mountain scenery, besides possessing a spring, the waters from which have few rivals and no superior as a curative agent in many of the most distressing complaints to which frail humanity is subjected. In cases of ulceration, scrofula, and debility from whatever cause, the cures said to have been effected by this water are marvelous. An analysis, as given by Col. WM. GILHAM, late professor of chemistry at the Virginia Military Institute, gave the following results, which are not only interesting to the physician, but to the scientific, as showing the materials of the rocks from which the spring derives its elements:

Carbonic acid	9.360	grains.
Sulphuric acid	53.383	"
Phosphoric acid	0.013	"
Magnesia	7.723	"
Lime	32.150	"
Oxide of iron	0.432	"
Alumina	1.729	"
Potassa	0.119	"
Soda	0.359	"
Chlorine	0.092	"
Organic extractive matter	3.733	"

These substances, existing in the water, give rise to carbonates, sulphates, phosphates, and chlorides, as follows:

Carbonate of lime.................... 8.642 grains.
    "   " magnesia.............. 1.389   "
    "   " protoxide of iron....... 0.617   "
Free carbonic acid................... 4.680   "
Sulphate of lime....................63.302   "
    "   " magnesia...............21.098   "
    "   " alumina................ 3.176   "
    "   " potassa................ 0.107   "
    "   " soda................... 0.750   "
Phosphate of lime................... 0.015   "
    "   " magnesia............... 0.011   "
Chloride of potassium............... 0.097   "
    "   " sodium................ 0.076   "
Organic extractive matter............ 3.733   "
Protoxide of iron....................Traces.

"This water contains an unusual amount of sulphuric acid."

There are many fine mineral springs in different parts of this county. So many indeed that the majority of them must always remain unknown to fame.

All of the above-named springs have telegraphic communications with all points.

TOWNS AND VILLAGES.

Christiansburg, near the center of the county, on the Atlantic, Mississippi and Ohio Railroad, is a place of about one thousand or twelve hundred inhabitants, including the suburb at the depot, which has a special name and a separate post-office called Bangs.

Here is situated the county court-house, containing the records of a great part of Southwestern Virginia. There are quite a number of hotels, stores, and establishments for the

manufactory of saddlery, tin ware, boots and shoes, etc.
Here is located the office of an enterprising and excellent
journal, *The Montgomery Messenger*, devoted to the improvement of its section. There are churches of the various
denominations, and schools both public and private.

The female schools are somewhat noted throughout this
region as being organized on a higher basis than is usual in
country towns.

BLACKSBURG.

This village has already been mentioned as being situated
toward the eastern end of the valley of Blacksburg. It is
a place of about two hundred and fifty inhabitants, and is
now noted as being the location of the Blacksburg Agricultural and Mechanical College, an unfortunate institution,
the child of our good old mother, Virginia, during the time
of the great prostration, consequent upon the event of the
" late unpleasantness," and could not, reasonably, be a very
vigorous institution.

This village is pleasantly situated, and would of itself
form a delightful summer resort, surrounded on all sides
with rolling grass and grain fields, dotted here and there
with handsome residences.

CENTRAL DEPOT

Is situated on New River, toward the western limit of
the county. This point is one of much importance in the
county, as having a great part of the repair shops of the
Atlantic, Mississippi and Ohio Railroad. Here is a large
round-house, and a lay-over station for engines, engine-men,
conductors, etc. It of course affords a very nice little market for much country produce. There are churches, stores,
hotels, etc.

ALLEGHANY STATION,

On the Atlantic, Mississippi and Ohio Railroad, is a small place, as you approach the eastern side of the county. It is situated in quite a tobacco producing section, and is of some importance as a shipper of that article. It likewise has stores and a church close by. It is the landing-place of visitors to the Alleghany Springs. Near to this is an opening on the zinc and lead vein.

BIG SPRING.

This place takes its name from a limestone spring of extraordinary volume, which rises close by, and flows into the South Fork of Roanoke River, upon the south side of which stream Big Spring (the station on the Atlantic, Mississippi and Ohio Railroad) is situated. This place is very attractive to persons seeking an excellent summer resort. The clear water, fine air, noble sugar-trees, and high mountains, with splendid fare at the hotel, render it very inviting.

There are several stores, and a factory for making lathes, upon which to turn wagon, carriage, and buggy spokes. The enterprising proprietor and inventor of this machine, Mr. Coffee, takes great pride in showing the machine at work, turning out two spokes per minute.

LAFAYETTE

Is an ancient village at the junction of the North and South Forks of Roanoke River, close to the eastern boundary line. It has one or two stores, a church near, and a fine flouring-mill.

PUBLIC SCHOOLS.

According to the recent report of DR. RUFFNER, Superintendent of Public Education, the public schools of Montgomery are improving. There are thirty-nine white schools, and nine colored. For these there are thirty-nine white teachers and nine colored teachers. Upon these schools there are in daily attendance about eight hundred white children and two hundred and fifty colored.

At Christiansburg and Bangs there are two graded schools, one for white, the other for colored students. The white school is one of six grades, the colored of two. The average daily attendance of the white school is twenty; of the colored, one hundred and twenty.

PRODUCTION OF CATTLE, SHEEP, WHEAT, CORN, AND TOBACCO.

This county is divided into several sections, of totally different agricultural capacity, by the different geological formations.

Those parts which are best adapted for cattle are of course the great grass areas, common to the limestone valleys. We have already seen that a great part of the county is of this nature. It is more than one half, when you add to the limestone grass lands those lands along Brush Creek which have been rendered fertile by the decomposition of felspathic rocks. Sheep can be raised profitably all over the county. Wheat is apparently in its native element. Corn does well on all the grass lands, and tobacco is suited to nearly all the soils in the county. It is mostly cultivated in the southeastern and northeastern parts of the county.

Of cattle, averaging the last few years, the county has shipped per year ....1,800 head.
Of sheep, " " " " " ....25 car-loads.
Of wheat, " " " " " ...35,700 bushels.
Of corn, " " " " " ....3,600 "
Of tobacco, " " " " " ..615,000 pounds.

TIMBER.

The different kinds of timber native to this latitude are very abundant in the northern side, and more mountainous districts of the southern belt of the country. The oak is predominant. Over much of the area between Brush and Gap Mountains, there are fine bodies of good hard wood for charcoal purposes. The south side of the county occasionally shows considerable bodies of hemlock and white pine, as well as good areas here and there for coaling.

WATER POWER.

There is much unused water-power in the county. New River, discharging about eighteen hundred feet per second at low stages, has fall enough at two or three places to be used; but on account of the height of the floods it might not be profitable to use it. Little River, discharging one hundred and eighty feet per second at low stages, will be a very useful stream. South and North Forks of Roanoke River, of rapid descent, present many locations suitable for wool-carding machines and grist-mills. Tom's Creek, and several other streams of smaller volume, afford constant streams throughout the summer. One of the advantages possessed by these rivers and creeks is the never-failing character of their flow.

GRAPE CULTURE.

The cultivation of any imported varieties of grape upon this elevated dividing ridge, between eastern and western waters, must be attended with much uncertainty. Experience has proven that careful attention paid to native varieties not only causes them to bear a grape of fine flavor, both for wine-making and eating, but an unfailing crop may, with greater certainty, be looked for with the native than with the foreign varieties.

BEE CULTURE.

The luxuriant growth of flowering trees, shrubs, and plants of this section would render bee culture, under proper man-

agement, a paying industry. There is much interest mani-
fested in the recent improvement in gums, but very little
more honey is made than is necessary for home use.

FISH CULTURE.

The work done in the last year or two, both by the State
Commissioner and private individuals, is beginning to show
in the large number of black bass, and other fine varieties,
making their appearance in New River and tributaries in
this county. In the Roanoke, near Big Spring, Alleghany
Springs, North Fork, etc., the large increase of fish has been
owing, to some extent, to the exertions of Captain Sumter, of
Big Spring.

PULASKI COUNTY.

It is difficult to find language that will introduce, with proper and merited description, each county as it presents itself, without drifting into a condition of sameness desirable to be avoided in treating of important matter. And when, as in the case of the different counties of Southwestern Virginia, candor would compel the impartial writer to use many adjectives of praise in speaking of their resources, the task of properly presenting the subject becomes still more difficult, if the appearance of mere flattery and adulation is also to be avoided.

Particularly is this the case with the county of Pulaski. The natural features of this county are nearly all of that order, which, if known universally, would fix the attention of the least observant, whether it was invited to the scenery, made up of mountains, forests, and broad streams, great grass fields, dotted with herds of fine cattle, or those extraordinary exhibits of mineral material and mineral waters that mark the belt in which Pulaski is situated. Though small in the extent of its territory, in comparison with the other counties of this section, Pulaski is making a wide and favorable reputation for the almost fabulous quantity of its iron ores and the extent of its fine coal fields, to say nothing of the ores of zinc, lead, and other minerals. In addition to which, it is making giant strides to obtain a position as the leading manufacturing county! Look at the large zinc-reducing establishment recently erected at Martin's Depot, and at the very extensive iron furnaces in course of preparation at the same place! Altogether, Pulaski has no reason to blush for the part it is performing, either for itself or for the State. The course it is now pursuing makes of it a most valuable factor in the final settlement of the disturbed financial condition of the State, enhancing, in large measure, the tax-paying power of its own and neighboring communities.

Pulaski is bounded north by the county of Giles and a small part of Bland, marked by the crest of Walker's Mountain; east, by New River, up to the mouth of Little River, which stream then forms the boundary for eight or ten miles on the southeast; on the south, by Floyd County, and a part of Carroll; west, by the county of Wythe. The south boundary line follows very much the crest of the Poplar Camp, or Iron Mountain.

HOW WATERED.

The county is well watered by New River, Little River, and their tributaries. Among the latter, Back Creek, Peak Creek, Big and Little Reed Island Creeks, and Laurel are the most important.

Geologically.—Pulaski shows nearly all the strata from the Potsdam to the Proto-Carboniferous inclusive. It is similar to Montgomery County in this respect, except that, in the elevation made by Draper's Mountain, it has a much larger exhibition of Potsdam rocks and ores.

The county is divided into four main geological divisions, in part owing to the position of Draper's Mountain, which severs the western half of the county nearly in the middle from west to east. The first, or southern section, is a great synclinal trough lying between Iron Mountain on the south, and Draper's Mountain, as above described, with New River occupying its greatest depression. The second, northern, is the Robinson Tract and upper Peak Creek country, a broken anthracite coal basin, bounded south by a fault line between the underlying Devonian rocks, and the upthrow of the Potsdam, and north by Little Walker's Mountain. The third, or eastern division, is the great plateau of beautiful grass lands, bedded upon Silurian limestone, occupying the space from

Cloyd's Mountain, southwardly, to the foot hills of Mac's Mountain, Draper's Mountain having come suddenly to an end in Peak Knob, leaving a wide plateau of grazing and farming lands between its eastern end and the line of New River, after that stream turns a northward course. The fourth division is the valley of Devonian strata lying between Little and Big Walker's Mountains.

It being somewhat unusual to find the Potsdam and Devonian rocks thrown into contact in this section of country, the accompanying geological section will be found to be located across Draper's Mountain, the north side of which shows this unusual occurrence :

DESCRIPTION OF SECTION.

The southern end of the geological cross-section on the southern boundary of the county, shown on page 41, begins about the crest of the main Iron Mountain, near a point where Mac's Mountain, a lateral spur of the main range, diverges from it on the north side.

These rocks represent the series about the division between the Huronian and Cambrian, or Lower Silurian. The conformability of the strata has been greatly disturbed, both by the pressure from the southeast, and by a great end-pressure or strain at right angles to the southeast pressure. Consequently, the dip is very variable. It may be recorded as being southwardly at high angles. To the south we have the hydro-mica slates, overlaid with a broad felspathic and quartoze series, terminating in a band of iron ore ; above this, nearly one thousand feet of Potsdam sandstone, with occasional bands of slates, some of them very dark ; over this, in the order of stratification, are two veins of brown iron ores, 6 and 9 feet respectively, separated by a band of slate not over 20 feet thick ; next to these come nearly 1,600

feet of Calciferous red shales and slates interstratified with
limestones more or less magnesian, and sometimes highly
ferruginous, here and there showing lead and zinc, the red
slates, etc., very frequently yielding fine specular iron ore.
After this an unknown thickness of the upper Calciferous
limestones, much folded and repeated as you approach the
basin of the river, where they begin to assume a more regular
and horizontal attitude, as often showing a gentle inclination
to the east or west as any other way.

From the river to Draper's Valley, which lies just south of
and parallel with Draper's Mountain, there is a repetition of
the red slates and shales, alternating with limestone, and
sometimes hard sandstone, so folded as to afford little chance
of ascertaining thicknesses. Draper's Valley is a beautiful
and fertile limestone valley, showing some ledges of limestone which make a beautiful ornamental stone belonging to
the Calciferous series. You then encounter Draper's Mountain, with the Potsdam rocks dipping southwardly at an
angle of about 50°, bedded upon buff and various colored
slates. A part of this mountain is a broken anticlinal—particularly that about Martin's Depot—with its northern limit
defined by the great fault just now mentioned, in which the
Potsdam and Devonian rocks are brought together.

The rocks of the Potsdam lining this great fault on its
south side are those which hold the heavier bands of iron
ore, and they are so thrown together for several miles
along this line, that the fault is marked by one of the heaviest beds of iron ore in Virginia. Passing northwardly the
section reveals about 2,500 feet of Devonian rocks, mostly
fossiliferous slates, dipping northwardly, overlaid by the lowest beds of the Proto-Carboniferous, showing some coal in
thin seams. You may then say you are in the Pulaski coal
basin until you reach Little Walker's Mountain, six miles to
the north. About Martin's Depot the red slates, which over-

lie the coal at some distance, are easily distinguished in a railroad cut, and in the bed of Peak Creek. From there until you reach Robinson's Tract, which lies next to Little Walker's Mountain, the stratification is much disturbed ; and Tract Mountain, which bounds Robinson's Tract on the south, is a rather broken arch compressed from the sides at the springing line, showing an upheaval of the larger coal veins. Passing across the undetermined limestones of the Tract, you reach the more undisturbed strata containing the best coal veins in the south flank of Little Walker's Mountain, dipping at various angles up to 45° southwardly and south-eastwardly. Underlying the coal measures is a few feet of fine grindstone, followed by slates and sandstones 1,200 feet thick to the Olean conglomerate, which here outcrops nearly at the crest of Little Walker's Mountain, dipping 40° south-wardly. This measure is over 300 feet thick. Next under this is hard sandstone underlaid with an alternation of slates and sandstones, some highly fossiliferous and calcareous, for more than 1,500 feet down to the black slates and carbonaceous limestones which represent the Coal Oil rocks. Leaving these, you pass northwardly over the upturned edges of the Marcellus and Hamilton, the Oriskany, etc., the Clinton and Oneida sub-epochs, to the northern end of the section in Big Walker's Mountain, on the northern boundary of the county.

A section toward the eastern side of the county would be similar, except that the great central part of it would be an exhibit of Lower Silurian limestones, leaving out the southern rim of the coal basin, and the showing of upturned Potsdam rocks presented in Draper's Mountain.

IRON ORES.

In the continuation—north 70° east, and south 70° west—of the flanking ridges of the main Iron, or Poplar Camp Moun-

3

tain, on the north side, is a thick series of the Potsdam and
Calciferous rocks, as shown in the cross-section. Throughout its length in the county of Pulaski, perhaps it would be
a difficult task to approximate even the quantity of brown
iron ore it will yield. The Mac's Creek Furnace is built in
this range ; the old Laurel Creek Forge also used ores from it.
With the exception of occasional flexures and dislocations,
this Iron Ore is continuous throughout the length of the
county. At Mac's Creek, the nearest veins to the furnace,
are the two spoken of as being 6 and 9 feet respectively, in
walls of slate, and occurring about the division between the
Potsdam and Calciferous ; but under this, geologically, about
the beginning, or bottom, of the series of sandstones, marked
by the Scolithus Linearis, seems to be the largest deposit,
or vein. At one point, about five miles southwest of Snowville, the ore beds formerly used by the Laurel Creek Forge
give such dimensions as follow :

In 1,000 feet length of the vein, 180 feet of this length
is 940 feet wide, the remainder having a width of 200 feet.
The vein goes down nearly vertical to great depths, this
being on the crest of a high hill, and it may safely be estimated that the ore can be stripped to a depth of 300 feet.
This condition, of great thickness, is, without doubt, owing
to end pressure on the great mass of the stratification, causing either a lateral flexure and reduplication, or an interlacing
of fragments. To a depth of 150 feet this body of ore will
yield over 8,000,000 of tons of brown iron ore of good grade.
On the south side of this ridge is Laurel Creek, flowing
northeast, on which are situated good ore beds of high-
grade iron ore, much of it being derived from the decomposition of ferruginous limestones, the Lowest of Lower Silurian
limestones.

In this vicinity is, also, Redland Mountain, near Little
River, a good ore-bearing ridge, containing that class of ore

which stains the soil a deep red. These measures trend southwest through the Mac's Mountain and Mac's Creek region, giving the great and almost unlimited deposits upon the strength of which Mac's Creek Furnace was located. As you approach Reed Island Creek going southwest, the duplication of the mountain chain seems to cease somewhat, and at Flannigan's, or Graham & Robinson's lead mine on New River, the Potsdam and lower Calciferous rocks have subsided so much as to leave exposed large masses of the next succeeding series, the white silicious limestones, and blue wavy limestones, which, while they are distinctly Lead and Zinc-bearing rocks, also carry in places very heavy deposits of sulphuret of iron. The brown ores resulting from their decomposition form very extensive beds, as at Rich Hill, and those interesting deposits of pipe ores at Andrew Moore's, below, on New River.

A few miles up Little Reed Island Creek from its mouth, near the Pulaski and Wythe County line, there are the only evidences of the terrace epoch to be seen in this section of country, as is also mentioned in the description of Wythe County. This is in brown iron ore of a very high grade, in terraces about 60 feet above the present level of the creek. These terraces are, no doubt, owing to the creek, in past ages, cutting through the immense beds of sulphurets of iron and copper in Carroll County above; and its waters coming down heavily saturated with a solution of iron, while passing here through a more level country, became stationary long enough to make these singular depositions. While this is the main source of the iron, it is very probable that the decomposing rocks of No. 2 close by added largely to that derived from the creek. These ores are largely in use by Forney's Forge at Allisonia, near the mouth of Reed Island Creek, at High Rock Forge, and by Graham's new furnace farther up in Wythe County, as well as by the Boom Furnace in Pulaski. They

have proven to be not only of very high grade, but in very
great quantity. The next great deposit of brown iron ore is that
in the north of Draper's Mountain, along the line of fault just
now mentioned. At one mile and a half south of Martin's Depot,
on the Atlantic, Mississippi and Ohio Railroad, at what is
familiarly known as the Honaker Ore Bank, may be considered
the heaviest exhibition yet ascertained along the lead, which
runs with intermissions both ways for several miles.   At the
Honaker Bank a body of the ore was measured, and found to
give a length of 1,000 feet, 75 feet thickness or width, by an
average elevation of outcrop above water in the small branch
near of 125 feet.   Throwing out one half for intrusions of
quartz and clayey impurities, this body will yield 500,000 tons
of good ore, down to the level of water in the small branch.
This little stream marks the division between two large bodies of ore.   Perhaps the deposit west of it is equally as large
as that just now described. This ore is underlaid by the
felspathic series of the lowest Cambrian, or upper Huronian
epoch, and would show over it the Scolithus-bearing series ;
but the great fault by which these low measures have been
brought up so high has protruded the Scolithus series, so
that it has been denuded and carried away.

The analysis of the Honaker brown ore is as follows :

Iron 58.00, Phosphorus None.—DR. GENTH.

There are also surface exhibitions of brown ores in the
south flank of Little Walker's Mountain, very probably the
result of a decomposing vein of iron carbonate in the Coal
Measures.   Occasionally, in the hills among the limestone
grass lands, brown ores show.   They are generally very pure,
but as yet undeveloped.

The next locality of brown ores of any consequence is the
line of Oriskany rocks in the south flank of Big Walker's
Mountain.   The measure of the Oriskany in Pulaski is not

generally more than 18 feet. Occasionally this is nearly all brown iron ore of high grade; again it gives way to manganese ore, and very often it is nothing but a highly ferruginous sandstone. It is nearly 20 miles long in Pulaski County.

RED IRON ORE.

Specular ore may, now and then, be found in the large Potsdam vein above described, but not in well-ascertained quantities. In the overlying red shales and slates it is in very considerable quantity, but not yet found well enough in hand to justify mining. It forms a very large proportion of these rocks, but unfortunately too generally distributed. It is only after their decomposition that it begins to enter as an important factor into the question of iron ores. It then becomes the parent of some of the fine deposits found in the limestones at lower levels. In Draper's Mountain there is a thin vein of red ore; also in Big Walker's Mountain. This last is the fossil red, which, from the specimens so far examined, does not bid fair to be of any consequence.

IRON CARBONATE.

There is no doubt a vein of black band about 20 inches thick overlying the coal, but as yet not developed. In the rocks at the northern base of Little Walker's Mountain there is a measure of highly carbonaceous limestone, about 15 feet thick, which is also impregnated with iron to the extent of about ten per cent. This is a low form of iron carbonate. It may be regarded as a very valuable measure for furnace purposes. It is just about the position of the coal oil series.

MANGANESE ORES.

Manganese will be found a large constituent of all the Potsdam iron ores, except occasional lengths on the veins,

where there seems to be but little.   It rarely ever assumes
the character of a pure oxide of manganese in those veins in
Pulaski, but in the Oriskany ores of Walker's Mountain it
is an ore of very high grade.   Now and then, as at the Alum
Springs and Spur Branch, it bids fair to be found in sufficient
quantities to make it a heavy item of transportation.

COAL.

The discussion, in a public print, of a matter of so much
importance to this county as the Coal is, is entered upon
with some reluctance in this work.   To describe the Coal Field
correctly, and outline the proper course to be pursued for its
successful mining, is a delicate point.   It occupies the northern and northwestern portions of the county, extending from
New River on the east into Wythe County, 20 miles, on the
west.   The upthrow of Silurian limestones by which its width
is limited on the south is observable along Back Creek
until you reach the eastern limit of Robinson's Tract, when it
begins to assume the appearance of a basin with northern
and southern outcrops.   It is really an irregular basin, from
a line across the country just below Martin's, until you pass
into Wythe County on the west, occupying all the ground between the ridge just north of Draper's Mountain and Little
Walker's Mountain, six miles in width.   It is much broken
toward the middle by several nearly parallel ridges, such as
Tract Mountain and Chellokee Ridge.   The coal measures are
only considered, along here, reliable on the side next Little
Walker's Mountain, to which the Altoona Coal and Iron
Co. has built a narrow-gauge railway from Martin's Depot
on the ATLANTIC, MISSISSIPPI and OHIO RAILROAD.   Just at
the point where the Altoona Company has hitherto
mined the coal there is a very considerable disturbance
of the stratification composing the south flank of Lit-

tle Walker's Mountain; hence the measures here given were
taken a mile or so farther east, where no such disturbance existed. Throughout the whole extent of workable coal
there are considered to be two reliable veins : the underlying
vein measures 2 feet and 3 feet at different points; and
where there is no folding or sliding of strata, it is separated
by 15 feet of slate from a vein giving 4 feet of solid coal with
3 feet of looser coal over it, over which, at 25 feet, there is a
vein of soft coal 4 feet thick. The middle vein, as in the Altoona mine, sometimes assumes a thickness of 22 feet; but
this is no doubt owing to crowding of the strata from pressure. From the outcrop southwardly, before any possible
fault intervenes to cut off the coal, it is an average distance of
3,500 feet ; and if Robinson's Tract should be a true basin,
there would be nearly two miles' width, from north to south, of
the best workable veins, though about the middle of the
basin the coal would lie very deep. One mile length on the
veins—that is from northeast to southwest—by a width of
3,500 feet will yield about 4,100,000 tons of coal, that
approaches very nearly a true anthracite in character. The
Altoona coal mine will this year dispose of 47,000 tons of its
two varieties of coal.

As you approach the western end of the county, close to
the Wythe County line, the coal near the Atlantic, Mississippi and Ohio Railroad, or southern side of the basin,
becomes more regular, and for an area of several miles is
valuable. The best vein shows six feet thick, and assumes
its best character about fifty feet below water level. It
may be suggested, that the character of the coals of these
Pulaski measures would be found much superior below water
level. It is fair to assume, where they have been subjected to so much disturbance, and the coal near the outcrops has been exposed so long to the action of the elements,
that much of the carbonaceous material has been lost ; but

the same constituents in the coal below water level have had
no chance to be eliminated to such an extent, or rather, to
escape, and an analysis of that coal, say two hundred feet
below water level, is likely to be found much higher in carbon, and lower in percentage of ash. It might be of interest
to mention the names of special localities where these coals
have better developed; but as the veins are continuous, from
New River almost to the Wythe County line, along the south
flank of Little Walker's Mountain, it is not necessary.

LEAD AND ZINC.

Pulaski holds a part of the great Zinc and Lead Basin,
which is developed to so valuable an extent in Wythe
County, a few miles to the southwest. About two miles below the mouth of Reed Island Creek, on New River, there is
a great fragment of zinc and lead rocks, now being mined by
Flannigan, and Graham & Robinson. Both above and below
this point these strata are thrown out for some distance by
the obtrusion of the underlying red slates and shales; here,
however, it is likely to prove of value. The dip of the rock
is northwest 30 from the horizon. The greatest mass of
ore is found at the junction of the white silicious limestone
with the blue and white lamellated wavy limestone. When
last examined there were one hundred and fifty tons of ore on
the ground. The mine was opened by a tunnel three hundred and thirty feet long. There are occasional displays of
ore east of this point, but of no great consequence as yet.
Lead also shows in small particles in Proto-Carboniferous
rocks in Tract Mountain, and in Lower Helderberg rocks in
Big Walker's Mountain.

SILVER.

Silver is a constituent of pebbles found in a thin measure
of conglomerate, which lies about one hundred feet beneath

Cross Section thro. Pulaski Co.

Levels Above Tide

Big Walkers Mountain

Little Walkers or Cloyds Mtn.

Robinsons Tract

Tract Ridge

Cherokee Ridge

Atlantic, Miss & Ohio R. R.

Fault
Brawker Iron Ore 75 ft.

Drapers Mtn.

Scale of Miles

New River
Lead & Zinc
Iron

Iron

Iron Mtn.

Levels Above Tide

the Coal Measures, but is not supposed to be in very considerable quantity.

Limestones of all kinds exist in all but the more mountainous districts. The blue limestones of the valley are good for lime, and those which weather white or light drab are hydraulic.

There are many ledges of fine building stone both in the limes and sandstones. Along Peak Creek, in the lowest of the coal rocks, the sandstones could not be excelled anywhere for the beauty and size of the stones, which can be obtained homogeneous in color and texture, breaking with a regular fracture, and weathering well; they are much sought after.

In Draper's Mountain and in Poplar Camp Mountain there are ledges which will yield rock suitable for furnace lining. In Draper's Valley there are ledges of limestones so variegated in color as to approach marble in beauty.

The Pulaski Alum Springs on Little Walker's Creek is now the only regularly kept watering-place in the county. In its neighborhood visitors are entertained by citizens also. The waters of these springs flow from Marcellus slates. They are distinctly alum and chalybeate, and are highly tonic in their character. The mass made from the water sells readily, and is now the source of some profit to the owners of the springs.

The valuable trees of Pulaski are white oak, chestnut oak, chestnut, poplar, walnut, hickory, white pine, and a few others, such as sugar-maple and buckeye. White oak seems to be the predominant feature in timber. In the valley between Little and Big Walker's Mountains there are large quantities of fine white oak.

WATER POWER.

Leaving out New River, Little River and Big Reed Island Creek are capable of supplying a very great water power. These streams each discharge over one hundred and eighty cubic feet per second. Little Reed Island Creek, somewhat smaller, yields fine power. Peak Creek, from its average fall being over twelve feet per mile, offers many mill sites. Back Creek also, though a smaller stream, is useful in that way.

MANUFACTURES.

If the metal-reducing establishments now built and in course of preparation be included under the head of manufactures, Pulaski occupies a prominent position among her sister counties. At Martin's Depot there is a system of furnaces recently completed, now reducing the zinc ores of the New River basin to metallic zinc, or spelter, as it is called. There are now two principal furnaces with about twelve tiers of retorts each, with an average daily capacity of about four thousand pounds together. To these are to be added soon as many more. On Mae's Creek, in the southern district of the county, there is a finely constructed iron furnace —the Radford Furnace—having an average daily capacity of twenty tons, now in operation. At Allisonia, Reed Island Creek, Forney's Forge turns out twelve tons of fine blooms per week. On this creek, just above, is the newly erected "Boom Furnace." At Snowville, on Little River, there is a good woolen factory for making kerseys, jeans, cassimeres, blankets, etc., but not now in operation. There are sufficient mills in the county to meet the wants of the people, some of them very good.

AGRICULTURE.

It is almost impossible to do justice to the subject of agriculture in the county of Pulaski. For so small a county, comparatively, it has established for itself an enviable name as a grass county. Back Creek and sections of New River, with occasional areas on Peak Creek and in Draper's Valley, afford some of the finest grazing lands in Virginia. There is much of the county, besides, devoted to mere farming operations, much of it being well adapted to tobacco, such as the hill lands, toward the southern side of the county.

SCENERY.

The rivers with its high cliffs, its sweeping bends, and clear waters, makes the most striking scenery of the county. Oftentimes, bordered as it is about the Horse Shoe with luxuriant and spreading grass fields on one side, and cliffs on the opposite side, it affords scenery of a very high order.

FRUITS.

All the fruits of this latitude do well in Pulaski. *Grapes* when properly cultivated and pruned yield a very certain return.

Fee culture is becoming gradually of more importance each year. The culture of fish by the State authorities and private parties has been looked after with a great deal of interest. The lake at Martin's is now stocked with black bass, placed there several years since by Captain Sumter, and New River is now affording, about New River Station, capital sport in the improved varieties, which are becoming somewhat numerous. Little River is likely to be used by the authorities as an excellent breeding ground for favorite fish, on account of the existence of a spring, which comes out near its west bank, in a very large stream from a deep cavern in the limestone rocks. In this place large quantities of catfish annually hibernate.

TRADE IN CATTLE, SHEEP, WHEAT, AND TOBACCO.

The annual shipment of cattle from Pulaski is about 3,800, of which one half goes to the English market. Of sheep there are exported annually about 6,000 head. There are about 15,000 pounds of wool used in the county and sold out of it annually; 68,000 bushels of wheat shipped; 5,000 bushels of corn annually shipped, and about 135,000 pounds of tobacco.

LINES OF TRANSPORTATION.

This county is directly on the great through line of railway from the Atlantic seaboard, through Tennessee, to the West and South. The ATLANTIC, MISSISSIPPI and OHIO RAILROAD passes through the heart of the county from east to west. New River, besides being the line of a navigation improvement now under the auspices of the general government, is also the line of the NEW RIVER RAILROAD, a road now being constructed. This is also one of the proposed lines of the PITTSBURGH SOUTHERN RAILWAY.

The county is well supplied with good country roads; one of them is a fine macadamized turnpike.

TOWNS AND VILLAGES.

Newbern, the county site, is a small town having the usual number of stores, hotels, and various kinds of repairing shops. It has churches of various denominations, and schools. At this place is published an ably edited weekly newspaper,

worth more than all of the other institutions of the town put
together. Dublin is two miles north from Newbern, on the
Atlantic, Mississippi and Ohio Railroad. It is a place of
several hundred inhabitants, containing churches, stores, and
schools. Martin's is now the place of progress in the county.
With its Zinc Works and fine Iron Furnace, soon to be completed, its railroad connecting the heavy beds of coal with
the larger railroad, together with its romantic location upon
a beautiful mountain stream, all bid fair soon to make it the
most considerable place in Southwestern Virginia. New
River Station will be a place of importance also, as the point
of junction between the ATLANTIC, MISSISSIPPI and OHIO RAIL-
ROAD, the NEW RIVER RAILROAD, and the NEW RIVER IMPROVE-
MENT. Snowville, on Little River, is an enterprising village
with a woolen factory, grist and sawmill, good stores,
schools, smith-shops, etc.

PUBLIC SCHOOLS.

The public schools of Pulaski have fared like the other
schools in Virginia the last few years; but are now decidedly
improved, and are, no doubt, on a permanent basis of prosperity.

It will not do to dismiss Pulaski County without saying it
is now entering, as a most important factor, into the question
of the future prosperity of the State. It would be well for
the example of those men, who have been so instrumental in
bringing about this healthy condition of affairs in Pulaski, to
be more widely copied in the different counties. If such
should be the case, how soon would the really fine resources
of the State be the means of her redemption from the embarrassments that have held her bound in the past.

WYTHE COUNTY.

To open the chapter on Wythe County in a manner worthy of the high claims it has upon the consideration of the public, will be quite as difficult as it will be to close it with the consciousness of having done justice to the subject.

Every endeavor has been made to treat all the territory described in this volume with the utmost impartiality; and it is not to be supposed, because development has been pushed to a much greater extent in Wythe than in any other county, that an impartial description of it, as it presents itself, is intended in the least to detract from the just merits of the other fine counties composing Southwestern Virginia.

On the contrary, an apology is due to the patient friends, who have so kindly awaited the appearance of this book, for the imperfect manner in which this great county, with its varied resources of a superior character, is treated. But it is due, also, to the reader to say that much that is interesting with regard to Wythe would have been omitted had the book made its appearance twelve months ago.

In the development of its different ore fields the county is making rapid strides toward a position of commercial importance, well calculated to excite the just pride of her citizens, as well as to encourage the friends of the State in the hope that so progressive a spirit, showing in other counties of the section as well, will tend far toward the early solution of those financial difficulties which have well nigh compromised her honor, and which, without the active development of the lately-hidden resources of the State, would find but a tardy settlement.

That Wythe is nobly doing its duty in increasing the tax-paying power of its own and neighboring communities, no one can doubt who will look at the different furnaces and mines recently put in operation in the county. And these

works, it may be submitted, supplying extensive home markets, besides employing the industrious labor of the country
at remunerative wages, are making the burdens of the State
much easier to be borne than formerly ; not only releasing
old residents from embarrassment, but bringing in new men,
wealthy, and, at the same time, willing to help the country
out of its troubles.

It is in no wise intended in such an introduction to under-
rate the importance of the agricultural interests of the county,
nor of the manufacturing enterprises which are struggling
through a healthy infancy to a mature age of great usefulness
and importance.

If conditions of transportation could once be made to assume a correct relation to the different interests of agriculture, mines, and manufactures, Wythe would not be long in
taking a leading position among the counties most noted for
high commercial prosperity ; and this comparison might very
safely be extended to the most favored localities throughout
the whole country.

HOW BOUNDED.

Wythe is separated from Bland County, on the north, by
Big Walker's Mountain, except six miles of the northeastern
end of the line, which leaves Big Walker's Mountain, and cuts
over south to the top of Little Walker's, or Cloyd's Mountain.

On the south it is divided from the counties of Grayson,
to the southwest, and Carroll, to the southeast, by Iron
Mountain and its extension, known as Poplar Camp Mountain.   On the east is Pulaski County, and on the west is the
county of Smyth.

Within these boundaries may be considered to lie an extent and variety of mineral and agricultural lands which,
taken together, are unsurpassed by the same area anywhere
else in the United States.

Alternating with each other, in the south side of the county, are wonderful veins and deposits of Iron ores, Manganese ores, and Lead and Zinc ores of extraordinary purity. While in the northern half of the county fine magnetic and brown ores lie close to good workable veins of semi-bituminous and semi-anthracite coal. Lying between these great mineral belts, and interlaced with them, are fine blue-grass and farming lands of a high order, mineral springs not being uncommon.

HOW WATERED.

The county is well watered by New River (which flows through the southeastern portion of the county), and some of its principal tributaries, such as Cripple Creek and Reed Creek. These streams, with their many minor tributaries, leave but a small space of the whole area which is not thoroughly well watered, and, like all mountain streams of the section, are unfailing.

New River yields, at the Wythe Lead Mines, about 1,500 cubic feet per second. Reed Creek, watering the central and northern portions of the county, passes, at different points, from 30 up to 180 cubic feet per second; while Cripple Creek, watering the southwestern portion of the county, yields nearly as much, presenting much excellent water power throughout the county, as the descent is sufficient to give, on every two miles' length of the smaller streams, a fall of over twelve feet average; while New River, except the twelve feet at Pearce's Falls, shows an average fall per mile of about eight feet.

GEOLOGICAL.

Beginning on the south side of the county, and proceeding north in the description, Wythe County holds the rocks of nearly all the epochs, and their subdivisions between the

4

Huronian and Cambrian on the south, and the Proto-Carboniferous toward the north side of the county, and in nearly
all the eras represented by these rocks nature seems to have
expanded herself to the full in a lavish deposition of some of
the best and most useful ores.

As will be seen by examining the accompanying sections,
the Potsdam sandstones, and subjacent hydro-mica slates and
conglomerates, compose the Iron Mountain mainly. Only
here and there do the felspars, which are so common in
Grayson, to the south, assume any importance. This Iron
Mountain is then flanked on the north by the red Calciferous
slates, the Potsdam and Calciferous being usually separated
by an extraordinary band of brown iron ore and manganese
ores of great thickness and persistency. Next to the north
of the red slates are the great bands of variegated limestones,
holding dolomites (sometimes bedded on bands of silicious
limestones), which are the gangue of the unsurpassed lead
and zinc deposits now being so well developed in the New
River section.

In these rocks are also the deposits of iron sulphurets,
which, decomposing, have left such vast deposits of pure
brown iron ore.

Passing north of this line, crossing the line of Cripple
Creek and New River, there are found the upturned edges of
these same rocks again, but now dipping southwardly, as on
the south side they dipped northwardly for the most part, or
were so overturned as to dip southwardly in reverse order ;
altogether making of the south side of the county a great
trough-like basin, flanked on the south by Iron Mountain,
and on the north by Lick Mountain, and its continuation,
Draper's Mountain.

In Lick Mountain, in the center of the county, the Potsdam
rocks, with their peculiar fossil—the Scolithus—form a great
broken anticlinal, giving way on the north side to the band

of red Calciferous slates and shales, from which they are separated by the usual bands of iron and manganese ores. On
the north of this is the great band of Lower Silurian or upper
Calciferous limestones, with steep dips flanked on the north
of Wytheville by the brown sandstones and black slates of
Pine Ridge for the west half of the county, but by a repetition
of the same limestones further east in the direction of Max
Meadows. To the north of these the conditions materially
change, the persistency of strata from northeast to southwest being broken by the influence of the great cross flexures
and compressions common to the range of mountains just
north of this line—in the north boundary of the county.
Thus, passing north of Wytheville and west of Queen's Knob,
you encounter the upturned edges of Lower Silurian limestones—here and there showing magnetic and brown iron ores
and variegated marble—until you reach a great fault at the
south base of Little Walker's Mountain; there you are suddenly brought into contact with the rocks of the Proto-Carboniferous, holding good coal veins. But when you pass
east of Queen's Knob, for nearly the whole of that portion of
the county there has been no bringing up, on so large a scale,
of the Lower Silurian limestones. On the contrary, very valuable areas of the Lower Coal Measures still remain, as those
north of Clark's Summit and Max Meadows. To the north of
these, as in the Cove, is a band of the lower limestones again,
running up into the Hudson River and Clinton series in the
Cove Mountain, but flanked on the north by the great fault
just mentioned, at the south foot of Little Walker's Mountain, bringing in the Proto-Carboniferous dipping southwardly.

These measures are underlaid in the heart of Little Walker's
Mountain, by Devonian rocks, including the representative of
the Olean conglomerate, and these again are underlaid in
regular order by the slates and shales of the Marcellus, Ham-

Cross Section through Wythe County, Va., at Wytheville.

NOTE.—The above section is intended to show the general positions of different rock formations, &c. In many places there are elevations and folds, too various in different localities to be represented in every feature, without loading the book with quite a number of sections. But the overabove will give a very fair general idea of the arrangement of the rocks of different ages.

Thickness of Rocks in Wythe County, Va., and their foreign equivalents.

	FOREIGN
Oneida	1500 (?) Lower Llandovery
Medina	1000 ft Upper Llandovery
Clinton	60 ft Upper Llandovery
Niagara & Sa	300 ft Wenlock
L.H. & Oriskany	10 ft Ludlow
Corniferous	50 ft

		FOREIGN
Marcellus & H	300 ft	
Chemung	300 ft	Old Red Sandstone
Catskill	60 ft	
Coal Rocks	1000 ft	

NOTE—Nomenclature of Dana is used.

	FOREIGN
Potsdam	1500 ft Stiper Stones
Calciferous Slates	1000 ft Tremadoc, Skiddaw etc
Cal. Great Lime S.	6000 ft Sutherland Limes
Cal. Ophileta	300 ft
S. Pot. S-Pot. & Sl.	380 ft
Chazy	3000 ft Llandeilo Flags, &c
Birdseye	300 ft

ilton, Corniferous, and Oriskany series, to get to them passing
through the American equivalent of the Old Red Sandstone.
Next to the north of these, in regular order, dipping southwardly, are the rocks of the Upper Silurian Age, including
the Dyestone iron ore series, taking us to the Oneida grit in
the heart of Big Walker's Mountain—our north boundary
line for Wythe County.

COAL.

Wythe holds a very respectable area of the upper New
River series of coal measures. This coal is usually assigned
to the Proto-Carboniferous Epoch. It shows thirteen distinct
veins, and many of the fossils developed would lead one as
readily to place this coal in one epoch as another, except,
perhaps, the upper measures.

As may be seen in the accompanying map, the coal is confined to the northern side of the county, running along the
south base of Little Walker's Mountain, except a considerable area in the northeast side, which spreads southwardly,
extending to and just over the Atlantic, Mississippi and Ohio
Railroad; this condition being observable near to, as well
as about four miles east of Max Meadows Depot.

The veins of coal in the south flank of Little Walker's
Mountain run, almost continuously for twenty-four miles,
through that part of the county, with a general average dip
of 30°, trend north, 70° east.

These outcrops are then separated from those nearer the
Atlantic, Mississippi and Ohio Railroad by an uprising of lower
rocks, as in Cove Mountain and the little ridge just south of
it. Thus the area in which the openings have been made
near Max Meadows may be said to be about six miles long,
east and west, and two miles wide from north to south, at its
widest.

The veins in Little Walker's Mountain have been opened

at several points along their length in the county, notably at
Boyd's Mine, where the Stony Fork of Reed Creek cuts
through the mountain; at Asa Brown's; at two or three
places north of Rural Retreat Depot in the west end, and at
Brown's Coal Bank, in the Cove.

As already remarked, these veins number thirteen so far as
known. Beginning below is an eight-inch seam, bedded upon
a thin stratum of quartzeous black slate overlying a heavy
band of grindstone grit. Next above this seam are ten feet
of black slate and hard sandstone, the sandstone being the
foot wall of a three-feet vein of semi-bituminous coal of good
quality; overlying this, are twenty feet of alternate bands of
black and gray slates and shales; then twenty-one inches of
really excellent flaming bituminous coal, of which the analysis
appears below; then thirty feet of gray and black slates and
shales and thin bedded sandstones, leading up to a fourteen-
inch vein of good bituminous coal; then one hundred feet of
alternations of gray slates and thin sandstones, holding nine
veins of coal, no one of which is a foot thick. These measures are not constant, though the veins are continuous for
great distances. The disturbances to which they have been
subjected have caused them, in places (as at Stony Fork), to
sometimes assume a thickness of eight feet for short distances. This may be owing to end compressions, as well as
to folding from the opposite direction.

In the Max Meadows and Clark's Summit coal area, the
most important openings are those made by Joseph Crockett,
two and a half miles north of Max Meadows Depot, showing
coal eight feet thick; and by Draper and others, one mile
north of Clark's Summit, near the Pulaski line. This coal,
for the greater part of the area, is of a semi-bituminous variety, nearly approaching an anthracite. At the Draper and
Clark opening the coal dips southwardly, showing several
good veins, the best of which is six feet thick. These veins

occupy a basin, the southern rim of which is one quarter of a
mile south of the Atlantic, Mississippi and Ohio Railroad, just
east of Clark's Summit.

The analysis of the coal from Stony Fork is as follows, by
THOMAS EGLESTON :

Water...................................... 0.34
Volatile combustible matter 21.30
Fixed carbon...............................70.32
Sulphur 1.61
Ash.. 6.43

The coal, the analysis of which is just shown, from Little
Walker's Mountain, is much preferred by blacksmiths. It is
also used to some extent in grates in the town of Wytheville,
giving very general satisfaction.

The quantity so far taken from these veins does not exceed
twelve hundred tons. From the excellent coke this coal
makes, and its nearness to existing lines of through transportation, it should assume a much more important position than
it now occupies.

IRON.

Brown Iron Ore.

It is difficult to know where to begin the description of
what is commonly known as "brown hematite iron ore,"
there is so much of it, of a superior grade, in different parts
of the county. The first great continuous horizon of it, mentioned in describing the geology, occupies a position in the
north flank of Iron Mountain, near the division between the
Potsdam and Calciferous sub-epochs. It is in the continuation
of the same great belt described as existing in the north flank
of Pilot Mountain, in Montgomery County, and in the Laurel
Creek country in Pulaski County. In Wythe County, at
nearly any point you choose to inspect along the range, you
will find heavy masses of this brown ore, sometimes mixed

with manganiferous iron ore, and sometimes giving way to
pure manganese ore. This great vein often attains a thickness of more than one hundred feet, but does not yield as
fine a quality of ore as the deposits which lie nearer to the
lead and zinc horizon.

The quantity of ore which this immense vein will yield in
the county, on the Iron Mountain spurs, is far greater than
that of all other deposits combined; but, as a general rule,
contains a few hundredths more of phosphorus than will admit it as a strictly Bessemer ore, as may be seen from an inspection of the analysis here given by Mr. BRITTON, of ore
from this horizon:

ANALYSIS BY J. BLODGETT BRITTON.

Metallic iron	57.98
Insoluble silicious matter	.49
Sulphur	.19
Phosphorus	.29

This ore can be seen in large masses by following up Francis Mill Creek near to its head, upon which are now located
the Sayers and Oglesby Furnace, and the Noble Furnace.
These furnaces, however, as will be seen further on, derive
their ores from the purer horizon nearer the lead and zinc
lead. It has been remarked, in the description of Montgomery County, that this Iron Mountain or Pilot Mountain
deposit of brown ore is derived from the decomposition of
sulphurets; a decomposition, however, which seems to have
been carried to great depths. Within the Calciferous rocks,
and within nine hundred yards to the north of the vein just
described, there are two more deposits, more or less continuous, rarely ever exceeding twenty feet in thickness,
and very much the same kind of ore. This series of veins
extends, almost continuously, the whole length of the south
side of the county.

The Second Horizon of Brown Ores.

Going northwardly, the next great belt of brown ores, are those extending from near the mouth of Reed Island Creek up the New River (both sides) to the mouth of Cripple Creek; thence up that creek to its head, into Smyth County, whence they pass on to the southwest.

This truly great and valuable band of ores is the one lying about the Lead and Zinc Belt, and upon which are located the furnaces and forges chiefly. It is sometimes several miles in width, being governed in this by the gentleness or steepness of the dip of the great band of rocks holding the lead and zinc. This ore, as has been stated, is due to the decomposition of sulphureted ores; and the hills in which it is usually found, being from 80 to 200 feet above the neighboring water-courses, there is mining and stripping above the undecomposed ore for many years to come. This band of ore, as on the old furnace lands of David Graham, seems sometimes to occupy a fissure in the great limestone belt. This may very well be true. It is plain that there has been fissure action along that line. A close inspection of the Old Lead Mines shows that lead, zinc, and iron have all been so acted upon by the heat resulting from pressure as to have been fused and interjected into the gangue surrounding and above it.

Following the zone either way a short distance from the old lead mines, you see these same measures resume their original position as a stratification between well-defined ledges. On the Graham lands this fissure action is evident from the lode cutting across the trend of the strata. And it can be accounted for by recurring to the action of the forces engaged in folding the earth's crust thereabouts. The crumpling or folding action was evidently from southeast to northwest, and where fragments of the crust did not slide up upon each other in monoclinal shape, they proved " pieces of resistance " to each other, to such an extent that, in order to satisfy the general, conditions of the whole action, they were crushed and fused by the great pressure, bringing about fissure action in the attempt of fused matter, gases, etc., to break their way to the atmosphere above.   In the extract quoted on page 59, Sir Robert Mallet gives a very good general formula deduced from his reasoning on the subject of Volcanic Force and Energy : but in this district these calculations would have to be based, perhaps, upon somewhat different data than those assumed by him.   Not to be guilty of too great a diversion just here, it may be as well to say that the once-heated substance of the earth, in cooling, no doubt left a crust upon the outer surface, which, as the whole continued to cool inward, was left more or less unsustained, except by its own strength as a great arch. This arch or crust, not being able to sustain its own pressure, gave way in certain lines of fracture (some of which are represented by the trend of the Alleghanies and Blue Ridge Mountains), now represented by great fragments extruded and riding up upon each other, the force from the opposite direction (as the force of compression is supposed to have been equal in all directions upon the spherical arch) being compensated by cross flexures and the interlacing and intersliding of great fragments.   Hence, though it may appear impracticable to apply the formula given—which was, no doubt, based upon equal resistance to a pressure exerted from all sides alike—still the study of it may lead to a determination of the problem as to the probable depth at which the fusing took place in this instance.   To the general reader this departure may be of no interest whatever, but it is interesting to some individuals to inquire into many of these things which are puzzling the curious of our day and generation; and it may be appropriate to submit the conjecture, before dismissing the subject, that the great lines of fracture represented by the position of the Alleghanies, Blue Ridge, Rocky Mountains, and other mountains, have resulted from lines of vibration in the earth's crust, established in their direction by known forces, commencing, no doubt, with the first movements of the earth upon its axis, and gathering in intensity and definiteness, having been modified and somewhat controlled by the different forces of magnetism and gravitation which were exerted, from time to time, by other heavenly bodies.

SIR ROBT. MALLET ON VOLCANIC ENERGY.

In treating the subject of the immense geological formation holding the iron, lead, and zinc in Wythe and adjacent counties, it is thought wise to introduce the following interesting exposition regarding the probable history of the veins, as illustrated by a quotation from the elaborate paper of SIR ROBERT MALLET, F.R.S., upon kindred subjects.

In an excellent treatise on Volcanic Energy, by SIR ROBERT MALLET, F.R.S., etc., kindly loaned the author by Professor Francis Smith, of the University of Virginia, are found, not only the formula mentioned on the preceding page, but a remarkably clear and able exposition of the origin of volcanic force and energy. SIR ROBERT MALLET employs his trained mathematical reasoning and elaborate experiments with wonderful tact; and not only the writer, but numerous others would be delighted to see him engage in an investigation, to show whether the expansive force of the heat of the still-heated nucleus of the earth has anything to do in counteracting the pressure of the arched crust upon itself, which, without any such check, would at once, by its own gravitation, begin to be exerted with destructive effect; this expansive force itself being held in equilibrium by the nicely-proportioned weight of the superincumbent dome, having such a thickness and weight as would be required to suppress a dangerous excess of expansive energy from below. Such an investigation, in such able hands, might lead to a clear and incontrovertible showing of the thickness of the earth's crust, within a few thousand feet of the truth.

The author will present the formula, with the theorem upon which it is based, leaving to the reader to investigate the *whole* subject in Sir Robert's paper.

"If a curved surface (of the nature of a hollow shell or membrane) be in equilibrium when exposed to forces acting normally to the surface everywhere, then the normal pressure at any point is equal to the force in the direction of the surface (or shell) at that point multiplied into the sum of the reciprocals of the principal radii of curvature.

*   *   *   *   *   *   *   *

"83. Let P (Fig. 7) be the normal pressure upon the unit of surface (square inch or mile) cut from a pair of intersecting ribbons of the curved surface, as a, b, and c, d, at

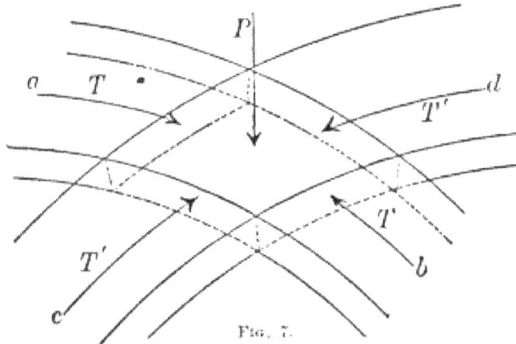

right angles to each other, and of unit breadth; T the tangential thrust on any of the faces of the unit square, respectively opposite (which, as being small in relation to the radii of curvature, may be considered as plane).

Fig. 7.

"Let the two radii of principal curvature (in a, b, and c, d,) be g_1 and g_2, then, as expressed in the theorem,

$$P = T \left(\frac{1}{g_1} + \frac{1}{g_2} \right), \quad \ldots \ldots \quad \text{I.}$$

"T having the same value.

"As regards the present application of the theorem, as the differences of g and g_2 for our globe are very small (comparable with the difference between the polar and equatorial radii), and scarcely sensibly affect the curvature of the surface within limited areas, we may consider our globe as spherical, and $g_1 = g_2$, whence Equation I. becomes

$$P = \frac{2T}{g}, \quad \ldots \ldots \ldots \quad \text{II.}$$

And

$$T = P \times \frac{g}{2}, \quad \ldots \ldots \ldots \quad \text{III."}$$

With this formula, which is sustained by such eminent authority as "Lagrange" and "Professor R. S. Ball," of

Dublin, Sir Robert calculates the thrust or pressure near the earth's surface, represented by T, to be 952,666 tons per square foot, basing the calculation upon the density and resistance of granite, which pressure is 472 times greater than is necessary to crush granite or porphyry.

For a continuation of the subject, the reader is referred to the paper by SIR ROBERT MALLET, recorded in the "Philosophical Transactions of the Royal Society," June 20th, 1872.

To return to the subject of the brown iron ore of the great belt of Wythe, it is of such importance, both from its purity and quantity, that it can very truthfully be declared worthy of a full discussion, both as to its origin and the accidents to which it has been subjected; but brief space renders it necessary to give all the information obtained in relation to it in as few words as possible.

Perhaps the largest development of these ores near the eastern, or Pulaski side of the county, will be found at Rich Hill, the property of Forney, and on Little Reed Island Creek, a mile or more south. No exact measure of these deposits is possible to be taken. That they are very extensive the wide surface showings amply prove; and at the beds in Little Reed Island Creek, the ores show for hundreds of feet in width, in terrace shape, about 60 feet above the present level of the creek—the only probable evidence of the terrace epoch in this section of the country. These ores have resulted, in all probability, not only from the decomposition of adjacent sulphurets, but the creek at a remote day has no doubt brought them down in solution from the great decomposing sulphureted beds and veins of iron and copper in the county of Carroll above. That they are in surprising masses, and in the position indicated, the most casual examination will show.

From these latter beds the "Boom Furnace" will derive the most of its ores when completed. The next good showing is at Graham's new furnace, south bank of New River. The next great outcrop then is at the old Peirce ore bank, about two-thirds of a mile south of Peirce's Falls on New River. A quantity of this ore was at one time used at the Old Poplar Camp Forge, now not in existence. This ore is evidently bedded on a limestone foot wall, which has a dip toward south-west of about 45 , and is distant from the uprising hard Pots-dam sandstone in Roaring Falls Mountain about one-fourth of a mile to the southwest. It appears to be over 40 feet thick at this point, trending east and west.

Then again, near the new furnace of the Wythe County Mining and Manufacturing Co., near Peirce's Falls, is a new opening in the ore of great richness and value. From both these places the furnace last mentioned will derive its ore.

A section taken north and south, across the general direc-tion of this series in the neighborhood, determines the ore field to be not less than two and three-quarter miles broad, which is divided nearly equally by New River at Peirce's Falls. Thus, on the north side of the river, nearly opposite the last-named furnace, is that part of this rich ore belt in which one of the first iron-masters of the section, Mr. David Graham, located his first furnace and the forge, rolling-mill and nail works, now known as Graham's Forge. Although a great deal of ore has been both stripped and mined in this neighborhood, it is highly probable that the great body of the ore still remains intact. A most cursory examination will show the great limestones, which accompany the lead and zinc veins, outcropping throughout this part of the ore belt. Their dip is usually southwardly through the Graham lands, but this is by no means constant. Holding as they do large quantities of sulphurets, their decomposition (sometimes *in situ*) has left large quantities of a pure brown iron ore. It is in this immediate section, near the residence of D. P. Graham, that twelve feet thickness of sulphuret may be noticed, just below water level, in a fissure between great masses of the limestone. This is a part of the fissure referred to on page 58. It is also alluded to in a paper on this section, reported in the "Transactions of the American Institute of Mining Engineers," vol. v., page 85, also in a report on the "Minerals of New River," made by the author to Col. Wm. P. Craighill of the U. S. Engineers, Exec. Doc., Nov. 25th, Third Session, 45th Congress, U. S.

Then, again, as you proceed southwest, in a section taken across this series of ores, near the old Wythe lead and zinc mines, the same broad exhibit of surface showing is to be found. Over the Lead Mine's Hill, south of New River, and in the vicinity of Walton Furnace, about two miles north of the river, these excellent brown ores have been mined and their intrinsic value fully tested.

It is possible that these ores make a metal that has no superior for car-wheels and all other purposes requiring an iron of uniform strength.

Then, again, going southwest, this series shows other great deposits in the vicinity of Brown Hill Furnace, Cripple Creek, the ores now mined there being taken from the great lead and zinc band near Abraham Painters, the deposits of iron ore being within 150 feet of the zinc ores.

In fact the original sulphureted veins are now known to alternate with the veins of lead and zinc, though generally distinct. This ore, which has been in use for some time at Brown Hill Furnace, assays as follows for Mr. JOHN M. SHERRARD, analyst :

Metallic iron.............................55.702
Silica................................... 4.590
Phosphorus................................ 0.0745

Then, one-half mile north of Brown Hill Furnace, the same

band of rocks outcrop, yielding very much the same kind of
ore.

A few miles west of the last-named vicinity the old forge
lands of the late Alexander Peirce present the same general
features, showing this brown iron ore in masses sufficient to
warrant the belief of its existence in very large quantities,
such as would be required to carry on operations of some
magnitude. In some places the mass of strata, which, by decomposing, has given such immense surface quantities of this
ore, is fully 180 feet thick, as found in position almost undisturbed.

Then, again, at the old Eagle Furnace, for miles on both
sides of Cripple Creek, these same conditions are observable.

The old Lockett and Huddle Forges, now out of use, have
sent some of the finest blooms to market, made from these
ores, that perhaps were ever made. Except for small quantities of impurities, other than sulphur and phosphorus, the
blooms were chemically pure.

A few miles farther west this series makes surface showings more southwardly. The Ravenscliff Furnace, the furnace of Sayers & Oglesby, the new furnaces on and near
Francis Mill Creek, and the Noble and Beverly furnaces derive and expect to obtain their ores from beds of this series,
which are well nigh thrust up within the area covered by
the Iron Mountain ores proper; but still they are decidedly distinct, the accompanying variegated limestones peculiar to the lead and zinc measures at once betraying their
origin.

Here these ores form a body which will yield not less than
one million of tons, together with its immediately neighboring deposits of the same series.

A close analysis rendered by Mr. JAMES AUMANN and
others gives nearly the following results :

BROWN HILL FURNACE, WYTHE CO., VA.

(P. 64.)

No. 1.—*Air Dry.*

Metallic iron . . . . .58.149 ⎫
Water . . . . .12.96 ⎪ Powder a deep reddish
Alumina. . . . . 2.32 ⎪ brown.
Silica. . . . . 1 09 ⎬ The roasted ore gave 67.235
Magnesia .Trace. ⎪ per cent. of metallic iron.
Phosphorus. . . . . .None. ⎭

No. 2.—*Air Dry.*

Metallic iron. . . . . .55.54 ⎫
Water. . . . . .11.72 ⎪
Alumina . . . . 5.89 ⎬ Roasted, the ore gave 63.04
Silica . . . . 2.73 ⎪ per cent. of metallic iron.
Magnesia .Trace. ⎪
Phosphorus. . . . . .None. ⎭

No. 3.—*Air Dry.*

Metallic iron. . . . . .47.82 ⎫
Alumina . . . . .11.53 ⎪
Water . . . . .10.83 ⎬ Roasted, the ore gave 53.28
Silica . . . . 9.87 ⎪ per cent. of metallic iron.
Magnesia. 0.06 ⎪
Phosphorus. . . . . .None. ⎭

Next to the southwest in this series occur the ores used by the old Porter Forge—of excellent character. Then, again, near to and just south of the Speedwell Furnace, these ores outcrop, across their general direction, for more than a mile. Their excellence and purity still remain a chief feature.

Then, again and again, as you proceed southwest, toward Smyth County, the same peculiar brown iron ore, either compact or honeycombed, makes its appearance, staining the soil red all along with a color peculiar to the ore.

It will be impossible to give any approximate idea of the quantity this great lead of iron ore will yield. That it will go high into the millions developments will no doubt amply prove.

5

The Brown Iron Ores and Manganiferous Ores of Lick and Draper's Mountains.

The ores of this series belong to the same horizon as those in the north spurs of the Iron Mountain and Pilot Mountain, and their general character is nearly the same, though they occasionally yield an ore of high purity.

Some of these veins and deposits are exceedingly massive, often running for great distances, preserving a thickness measuring sometimes one hundred and fifty feet, but usually not over twenty-eight or thirty feet, the fine manganiferous ore of the Glade Ore Bank not being over ten feet thick.

Both in Lick Mountain, on either side, and in Draper's Mountain, these ores are persistent, and give the following results by analysis:

No. 1.—Taken from a deposit of Kent & Stuart's, near A. Hoffman's, south of Wytheville; analyzed by Dr. J. W. MALLET, gave the following results:

Ferric oxide.............................77.46—Metallic iron, 54.43.
Manganese oxide............................. 0.64
Alumina.................................... 1.07
Lime... .32
Magnesia.................................... .45
Silica and insoluble silicious matter.............10.60
Phosphorus prot oxide......................... 0.27
Organic matter...............................Trace.
Water....................................... 8.72

No. 2, by the same chemist.......................46.62 p. ct. met. iron.
No. 3,   "   "   from the Sun Rocks.........55.61   "   "
No. 4,   "   "   Stroup's Branch...........46.85   "   "
No. 5,   "   "   6¼ miles south of Wytheville, 46.32   "   "
No. 6,   "   "   above Hudson's............59.16   "   "

These ores are all of the horizon in the lowest part of the Calciferous beds, and lower than the lead and zinc zone.

They may be said to extend in lines, not so persistent from the Wythe-Pulaski line, just south of Clark's Summit, through Draper's and Lick Mountains, in the direction of the White Rock or Panic Furnace, but dying out with Lick Mountain, before reaching the Wythe-Smyth boundary line. They again come up in the series of ridges making up Glade and White Rock Mountains in Smyth County, and are fully described in detail in the chapters on Smyth County.

Going northwardly, passing over some minor deposits in the Coal Measures, in the eastern end of the county, and a band of Oriskany brown ores in the south of Cove Mountain and Queen's Knob, the next deposit of magnitude is in a line of ores extending through Crockett's Cove, and found in position between the Black River and Trenton limestones, from six to eighteen feet in thickness. This range of ores extends from the western termination of Queen's Knob, through the above cove, nearly to the northeast corner of the county, yielding an ore generally regarded as being of good quality.

The Cove ore, just mentioned, together with the band of ores lying near the Coal Measures in Little Walker's Mountain, and the Oriskany brown ores and others in Big Walker's Mountain, make up quite a respectable aggregate for the north side of the county.

The Coal Measure ores seem to present two distinct bands : one lying just south of the measures throughout the whole length of the county, giving masses frequently 10 feet thick, and the other a vein 30 feet above the upper coal vein, about 18 inches, and, likewise, persistent. The analysis of the first by THOMAS EGLESTON, of Columbia College, is as follows :

Metallic iron 57.7
Silica . . . . · 1.56
Sulphur 0.085
Phosphorus 0.850

The analysis of the Coal Measure brown ore taken from Stony Fork Gap, is as follows, by the same chemist:

Metallic iron....................................54.00
Silica.. 8.71
Sulphur...................................... 0.104
Phosphorus................................. 0.205

After this, and last of any consequence, is the Oriskany brown ore in the south slope of Big Walker's Mountain. This measure is often 18 to 20 feet thick, sometimes perfectly free from silica, but usually well mixed with it; sometimes highly manganiferous, and occasionally replaced by manganese. Fragments of this ore taken from near where the Tazewell Turnpike crosses this ore series, give the following analysis by THOMAS EGLESTON:

Metallic iron 43.5
Silica..27.60
Sulphur 0.158
Phosphorus 0.082

This band of ore is generally persistent, and frequently along its length in Wythe County yields fine bodies of ore of both iron and manganese.

To conclude the chapter on brown iron ores, it might be appropriate to give some idea of the quantity the county of Wythe is likely to yield. An impartial examination into this question of quantity will leave any one at a loss to approximate the great array of figures necessary to determine it. Could these brown ores, the magnetic ores, and the red iron ores of the county be brought together, such would be the quantity, and such the general character, that it would be difficult to find anywhere else all the conditions so favorable for the production of a high grade of metal on a large scale.

Red Iron Ores.

Occasionally in the great New River and Cripple Creek series of brown ores, the ores will assume the character of a red or true hematite, but these instances are only local.

No doubt there are valuable bands of specular ore in Iron Mountain, but they have not been sufficiently explored yet to determine their exact position, measures, quality, etc.

But Wythe County, like Washington, and the west end of Smyth County, possesses a valuable band of semi-magnetic red iron ores, which sometimes attains a thickness of 15 feet, and rarely measures less than 9 feet.

What is known as the Yost, or Blair ore bank, found two miles northwest from Wytheville, is in this series. It yields as follows, by the analysis of THOMAS EGLESTON of Columbia College :

Metallic iron 61.7
Silica. 4.21
Sulphur 0.096
Phosphorus 0.075

Much of the ore in this band will do better than the above in every respect; and, as to quantity, so unusual has it been to attribute any large quantity of ore to the northern side of the county, by the old iron workers on the south side, that it would be difficult to convince the general public of the fine prospect that a good yield of ore may be expected from the beds under discussion. An examination of this line of ores, extending over a period of several years, shows that the ore near Wytheville is not an accidental deposit, but belongs to a series, composed sometimes of more than one vein, which belongs geologically to a system of rocks near what are known as the St. Peter sandstones, sometimes next to them,

and again separated from them by a broad band of black
slates and dark ferruginous sandstones. In Washington
County this condition is amply shown at the Gollaher ore
bank, four miles east of Abingdon, as well as at other points.

It is on the north side of Pine Ridge, in Wythe County,
that the best exhibits of this ore are found, and a farther
and closer examination, in the thirteen miles length on the
west side of the county, is expected to show several valuable
deposits.

The only other band of red iron ore in the county yet explored is the dyestone, or fossil band on the south flank of
Big Walker's Mountain, near the north boundary line of the
county. This ore is in two distinct measures—one, the fossil ore, 18 inches thick, made up of aggregations of fossils
and pebbles, yielding rarely over 45 per cent. of iron. The
other is an ore of compact slaty structure, which will yield
more iron no doubt, and would be a fine ore to mix with
other varieties. This latter is found three feet thick on the
road leading from Mount Airy to Rich Valley.

Magnetic Iron Ore.

Except in the red ores, two miles northwest from Wytheville, magnetic ore is not positively located in Wythe County
in any appreciable quantities. In the Iron Mountain a vein
of this ore is reported. This report is credited by a great
many iron men, but as yet it is not positively ascertained.

Sulphureted Iron Ores.

It is the belief of the author that all the ores enumerated,
except the dyestone ores, and the stratum of ore nearest the
coal measures, are derived from the decomposition of sulphureted ores. This opinion has been adopted after a long
and close study of all the deposits and their extensions at

many different points. But the quantity of iron sulphuret, known as such, in the county, is great.

In the lead and zinc veins, and in strata lying near to them, are large quantities of iron pyrites. This taken in connection with the quantity of sulphuret below water level, underlying the great iron veins mentioned, must aggregate so vast a quantity as to be without limit.

MANGANESE.

The localities in which manganese is found in any quantity are Iron Mountain, the Reed Island country, Lick and Draper's Mountains, the Glade, and Big Walker's Mountain.

The manganese ores in the north spurs of Iron Mountain are sometimes quite pure and abundant; usually as black oxide, but occasionally a handsome crystalline ore of high purity. The ores near New River, above the mouth of Reed Island Creek, belong to the variety known as Manganese Cream, and seem to be abundant.

At the Glade ore bank, four miles southwest from Max Meadows Depot, the manganese is found both combined with and entirely distinct from the iron ore. It appears to be in quantities.

In Lick and Draper's Mountains there are occasional deposits, the extent of which is not yet fully known. The ores found in Big Walker's Mountain are those mentioned as running with the Oriskany Rocks. They are now and then very pure and handsome.

LEAD AND ZINC.

It would be a pleasure to give a very thorough and complete description of the lead and zinc deposits of Wythe; their geological position, the kind of rocks in which they occur, and their exact thickness; together with a complete

history of their discovery and utilization. But it is feared
that many interesting points will be overlooked ; not through
any disposition to slight one of the most important and interesting subjects mentioned in the book, but want of space
compels brevity.

Geological.—The lower rocks of No. 2, in which occur the
great lead and zinc series of Southwestern Virginia, pass
almost continuously through the whole length of the southern section of Wythe County ; not always presenting the
strata of like richness in ore throughout, but over considerable areas, developing the veins in thick and massive measures. Here and there, the strata next in order beneath have
been thrust up so prominently across the direction of the
series, as to cause the extrusion and consequent loss of some
of the lead and zinc ores.

The rocks in which the ores occur are usually mentioned
as dolomites, but this statement cannot be fully accepted, as
much of the material accompanying the ore strata is a highly
crystalline limestone, sometimes silicious. It would be difficult to say precisely how far above the last of the Scolithus-
marked sandstones the lead and zinc zone lies. The two are
separated by alternations of red Calciferous slates, and blue
and white limestones, in some places nearly 900 yards thick,
with the immediate ore-bearing ledges generally, but not
always, resting upon silicious limestone. The ore-bearing
strata is marked all the way through by a wavy white and
blue spotted limestone, looking as though it was once full of
what now appears an indistinct fossil ; or, perhaps, owing its
appearance to gentle wave action in a shallow, chopping sea.
The ore strata along the continuation of the series in a direction north 70° east, and south 70° west, dips at various angles.
At the old Wythe Lead and Zinc Mines, at Austinville, New
River, the measures and dips are as follows : The principal
vein was found 40 feet thick, dipping south 20° east, at an

angle of about 70°, reached by a tunnel 1,600 feet long, the shaft from the top of the hill down to water level being about 245 feet in depth. The walls of the vein here are the so-called dolomite—highly crystalline. It is highly probable that this immediate portion of the series has been so acted upon by heat resulting from pressure that the original character of the walls has been greatly changed; but at Hendrick's, formerly Kitchen & Painters', the original conditions are preserved. Thus, according to the measures taken there, a true reading is as follows: Beginning on the floor or southeast wall of the main measure, we have 144 feet of heavy zinc blende-bearing strata dipping northwest at an angle of 30°; then 36 feet of dolomite with occasional spots of zinc and lead; 36 feet of iron sulphuret and oxide; 90 feet dolomite rock, containing large veins and deposits of zinc and lead sulphuret, one of which is 18 feet thick; 180 feet of iron, zinc, and barytes heavily disseminated in the rock; then toward the northern side or hanging wall, an indefinite amount of crystalline limestone more or less charged with barytes.

At the Bertha Zinc Mine, in the eastern or lower end of the county, and at Forney's Mine, next west of the Bertha, the mining has not extended below the decomposed surface ores, hence no exact measures have been taken there. At Sayers, on Little Reed Island Creek, lead sulphuret in some masses has been mined; and toward the west end of the series, on Upper Cripple Creek, on land of D. C. James and others, west of that point, lead sulphuret and fragments of zinc ore are occasionally found.

The kinds of zinc ores of the whole series are oxides, sulphides, and large quantities of a silico-carbonate—both at the old Wythe Mines, at Austinville, and at the Bertha and Forney's Mines. About 15,000 tons of zinc silicates and carbonates have been taken from surface mines at the Wythe Lead

and Zinc Mines, and from the Bertha Mines about 2,100 tons
of a very handsome silicate and carbonate, mixed with oxide,
and smelted at the new zinc works at Martin's Station,
Pulaski County, Virginia, yielding 30 per cent. of metal.

In the deep workings of the Wythe Lead and Zinc Mines
the zinc ore is a blende principally, as well as at the Crusen-
berry & Kitchen Mine. At the latter two the ores of zinc
and lead are sometimes alternately stratified. The great mass
of these ores in the county is far above the power of computation, as they are not only thick, but extend over much
ground. The kinds of lead ore are, galena, carbonate, oxide,
arseniate, phosphate, and occasionally molybdate; galena
being the most abundant ore, the carbonate is nearly exhausted, and the others exceedingly rare.

THE FURNACE AND WORKS OF THE WYTHE LEAD AND ZINC MINES
COMPANY.

The first work done at these mines was about the year
1756, the lead being reduced from the ore in the most primitive manner; but it was not until the mines fell into the
hands of Raper, Sr., and his associates, about 1830, that the
old Scotch system was introduced. This, with some modifications, is still in use. The power used to crush and separate
the ores and gangue is derived from New River, upon the
south bank of which the works are situated, and they now
employ in the mines, furnaces and separators about 150
men.

In the last three years the company has constructed four
round buddles, twelve jigs, and a new roasting reverberatory
furnace; having, also, introduced a fine air compressor to
ventilate the mines with, and to run a Burleigh drill, the air
being forced through a four-inch wrought-iron pipe.

The company sells its zinc ores, only reducing the lead.

The product of the mines is now three tons of metal daily. As the markets seem to justify it, this product is made into pig lead or shot, sometimes both. The shot tower in use is a shaft 242 feet deep, situated nearly at the farther extremity of the 1,600 feet tunnel. The shot is then loaded into a car on a tramway and run out to the shot-house at surface, where they are put through the process of separation, sizing, and glazing. The shot tower is, perhaps, one of the best in the world ; and, owing to the nature of the lead, the shot made is looked upon in the markets as the equal, if not the superior, of the best shot made anywhere else.

It may be as well to say, before closing the subject, that the capital stock of this company is $400,000, divided into shares of twenty dollars each.

COPPER.

Many of the most valuable ores being so lavishly distributed in different parts of the county, it would seem that it had already enough without attempting to enumerate others ; but so far as copper ore is known in Wythe, the quantity ascertained is small.

Beautiful specimens have been taken from ground near Abraham Painter's, by Gallimore, an experienced miner. They comprise both carbonates and sulphides, the former apparently resulting from the decomposition of the latter. It is not probable that the vein is a thick one. It occurs near the middle of an eighteen-feet measure of lead and zinc, mentioned in the description of the Lead and Zinc Zone.

The slates of the Calciferous sub-epoch sometimes yield small quantities of Copper ore, but the only other deposit worthy of mention is a body of ore yielding nine per cent. of copper, lying one mile northwest from Max Meadows Depot, which is supposed by many persons to have resulted

from glacial action. It is a sulphuret in course of decomposition, and may really be a larger deposit, and nearer its
parent vein than is generally supposed; for a long and
patient examination of the line of Lower Silurian limestones
on the north side of the Lick Mountain range, shows the
baryta beds, which pass through the greater part of Wythe
and Smyth counties, to be a copper-bearing series as well.
Occasionally, this series shows very handsome bodies of
copper ore; and it may be taken with almost absolute certainty, that the Max Meadow copper ores, as well as those
showing in Smyth County, near Mount Airy, result from
the disintegration of this series of rocks.

GYPSUM.

Hydrous sulphate of lime is found in beautiful crystals at
Stoner's, on Cripple Creek, near where the Grayson turnpike
crosses. The amount so far brought to the surface does not
warrant a conjecture as to the probable quantity. Future
developments may show much more than is now suspected.

MARBLE.

The different varieties of variegated marble and onyx-like
limestone in the county are very great. Handsome varieties
of brown and red variegated ledges are found in heavy
masses on the lands of Abraham Umbarger, three miles
northwest from Wytheville, supposed to be Bird's Eye and
Black River series. The fossils contained in them are, however, very indistinct, rendering it premature as yet to declare
the exact age of the rocks. From these quarries stones of
any size may be obtained. This series of rocks extends for
many miles through the county, being thrown out east, by
Queen's Knob, and passing westward into Smyth County,
but not always presenting the same beautiful appearance.

In Black Lick Township, near the residence of Mr. Davis, both the brown and nearly white varieties are found.

At Frye's Hill, five miles southwest from Wytheville, the variety resembling onyx is found in masses about a large cave, no doubt derived from the decomposition of the soluble limestones of which the hill is composed. Many of these blocks are like alabaster in purity, translucent, and much of it tinged with amber coloring.

This material is found at places northeast and southwest of Frye's, and at various places on Cripple Creek.

BARYTES.

Sulphate of baryta is found near Frye's marble quarry, not in very large quantities, but very pure. This is nearly in the true lead and zinc zone; but on that side of Lick Mountain the series is barren of both lead and zinc.

Barytes is found in very large masses on the north side of the great lead and zinc deposits. On a hill near Painter's Store, Brown Hill P. O., the quantity of this mineral is very great, extending along with the lead and zinc for miles. Other localities in the county require more development to be worthy of notice.

KAOLIN.

In Lick Mountain, as well as with the coal measures of Wythe, there are large quantities of kaolin. That accompanying the coal veins doesn't vitrify under great heat as does the Lick Mountain clay.

The kaolin of Lick Mountain shows to best advantage toward the head of Stroup's Branch, about five miles south of Wytheville. This deposit is in the Potsdam series, and is nearly 50 yards in thickness, showing a length of two miles.

TIMBER AND CHARCOAL.

Very few of the great timber boundaries, once so plentiful on the south side of the county, still remain intact. The great demand for charcoal for furnace purposes has thinned out the timber on that side of the county very much; but on the north side and in the middle section there are large boundaries of very good timber. The predominant kinds are white oak, chestnut oak, black oak, red oak, spanish oak, hickory, poplar, walnut, sugar-maple, chestnut, white pine, yellow pine, butternut, and all other trees and shrubs native to the climate. The most numerous is the white oak; white pine has been greatly decimated for shingling; but there remains a great deal of chestnut oak, good for tanning purposes.

AGRICULTURE.

In Wythe County agriculture has been carried to its highest perfection only in the department of grazing. The inexhaustible fertility of some of its soil, together with the care and economy habitual to many whose lands were not originally so fertile, enables the county to make an exhibit alike flattering to the land-owners, and worthy of the widest publicity.

It would be an invidious task to attempt to mention those whose labors have contributed most to place the county in the fine position it is esteemed to hold among the best grazing and farming counties in the State. The result of their efficient labors is shown in the fine herds of the best improved cattle and flocks of sheep, and the intelligent use by them of fertilizers in improving their wheat and corn lands is a compliment which is well earned.

That the whole state of agriculture in the county is making fine progress under all the discouragements it has had

to contend with, no impartial observer can doubt. It is
plain that the great improvement in this direction, together
with the rapid development of the different massive bands of
ores, is fast raising the county to a position among the fore-
most in point of wealth and importance. The cattle-men of
the county have been all the time abreast with the leading
men in like pursuits in others of the best grass countries in
Southwestern Virginia. And this is saying a great deal, for
the herds and herd books throughout the section named,
when once inspected, will show an advanced interest and results achieved, that would astonish both the English and those
Americans of other sections, who arrogate to themselves the
name of being first in such pursuits. It would be a pleasant
task indeed to follow this subject through all its details,
bringing in a description of the fine localities whose high
state of improvement lends such interest to the beautiful
landscapes, in which different parts of the county abound;
but a regard for time and space compels the abandonment of
a subject of the highest interest.

MANUFACTURES.

Beside the iron furnaces and forges and the lead works, the
different manufacturing enterprises of the county are few in
number, but seem to be directed by men who are determined
to win success.

At Wytheville, and in that vicinity, the plow works, found-
ries, machine shops, wagon and carriage-making establishments, and tanneries seem to be running to their utmost
capacity, some of them even finding it necessary to increase
their facilities for manufacturing. Thus, the best grades and
most improved patterns of plows, saw-mills, cane-mills, cast-
ings, wagons, buggies, carriages, leather, harness, and other
articles are made there and distributed over a large extent
of country.

At Kent's Mills the fine water-power of Reed Creek is utilized to run a good flouring-mill, besides a first-class woolen mill, which turns out cassimeres, linseys, jeans, blankets, etc.

At Max Meadows is a tobacco factory, engaged principally in the manufacture of chewing tobacco. At Crockett's, beside at Wytheville and other places in the county, plaster is ground for use as a fertilizer. Flouring-mills, saw-mills, and ordinary grist-mills abound throughout the county. Perhaps the greatest power in the State unused is that at Peirce's Falls, on New River.

SCENERY, ETC.

The mountain ranges, divided by valleys, which are threaded by numerous streams of widely different volume, flowing awhile through beautiful green meadows, and then under lofty cliffs of massive limestone, render the scenery very attractive.

There can be nothing more beautiful than that part of New River near the lead mines, or the fine stretch of river at Jackson's Ferry. At the latter, the southern bank rises abruptly into a cliff, which is crowned by an old tower covered with ivy ; while the northern bank slopes away gently for several hundred yards to an eminence, upon which is an elegant country residence, built in a style of architecture in perfect keeping with the noble scenery around it.

At Wytheville, the north side of Lick Mountain, in summer afternoons, while the slanting rays of the sun bring out in fine relief the prominent ridges and deep hollows with their marvelous alternations of light and shade, presents a picture of incomparable beauty.

At Max Meadows, the broad fields, whether decked in the fresh colors of leafy June, or crowned with the golden yellow

of the fall season, contribute many rare views, to which the
streams and neighboring mountains lend a character vivid
and animating in a high degree.

SCENERY, MINERAL SPRINGS, ETC.

There are many such views, which the feeble power shown
in this volume could but faintly portray.

The Chimney Rocks, near Wytheville, have been for many
years a resort of the young people in search of the commanding views, cool shades, and the fine limpid spring near that
elevated place.

The great body of the valleys of the county being composed of heavy bands of limestone, their unequal solubility
has caused many caves in the county, some of which, if thoroughly explored, would rival the Luray Cave in extent and
beauty of adornment. The section of the county known as
The Cove, among many other localities in the county, seems
to have some of great beauty, besides being of easy accessibility.

The mineral springs of the county are confined principally
to the line of the coal rocks and the adjacent strata, except
the fine alum-chalybeate, which is now brought into Wytheville through pipes. The spring from which this water is
derived, like several others near it, is in the black slates of
Pine Ridge, whose position is near the St. Peter sandstones.
There are few except alum-chalybeate springs in the county,
of a strictly mineral character; and these, while of high
medicinal virtue, are numerous.

WYTHEVILLE.

Wytheville, the county seat, which, by the last census, contains about 2,300 inhabitants, is situated near the center of

6

the county, on the Atlantic, Mississippi and Ohio Railroad.
The high improvement of its streets, the air of neatness and
cleanliness, and its elevation of 2,300 feet above the sea,
combine to render it a very attractive place in the summer
season.

It out-rivals a great many of the lesser watering-places in
the number of visitors, in search of health, it entertains during the summer and fall. Beside several well-kept hotels
and boarding-houses, usually dispensing well-cooked and
wholesome fare, the fine alum and chalybeate water, now
flowing from hydrants on the streets, adds an attractive and
valuable feature.

It is hardly necessary to say that all denominations of
Christians, common to other parts of Virginia, have churches
in the town or its vicinity. The schools, though unpreten-
tious, are among the best in the State. Beside well-regu-
lated public schools, there are private boarding-schools, of a
select order, for young ladies.

In addition to flourishing manufactories and machine shops,
for the making of anything from an engine to a plow point,
there are numerous stores which are supplied with a more
than usually select list of goods. Thus, everything is found
for sale in ordinary merchandise, millinery, fancy goods, no-
tions, groceries, hardware, leather supplies, and tinware.

Though sewing machines, reapers, and mowers, and the like
are not made there, several companies are represented by
handsome displays. The carriages, buggies, and wagons
made at its factories, and proportioned, in strength of make,
to the roads over which they are expected to run, out-rival
any others for use in the adjacent country.

The two newspapers, *The Enterprise* and *Dispatch*, published at the place, always occupying an advanced position
in advocacy of progress and improvement, give full information of the proceedings of all the courts, including the Su-

preme Court of Appeals of the State, which sits there annually, beginning about the 10th of July.

Wytheville is healthy, and the scenery around it is of a
high order.

Max Meadows, a name now commonly applied to the depot
of that name on the Atlantic, Mississippi and Ohio Railroad, was originally used to designate the wide-spreading
meadows of great fertility in which the depot was built. The
place has several excellent stores, and a tobacco factory. Max
Meadows is one of the great shipping depots for pig iron,
lead, and zinc.

Rural Retreat, in the west of the county, besides being a
depot on the Atlantic, Mississippi and Ohio Railroad, is a
growing place, having an advanced school, good stores, shops,
etc.

Crockett's, is a depot on the Atlantic, Mississippi and Ohio
Railroad, seven miles west of Wytheville, and is now the
shipping point for much of the pig metal made in that end
of the county. It has a good steam mill for grinding plaster,
besides other improvements, stores, etc.

There are quite a number of other noted places in the
county, but their description is now impracticable for want
of space.

LINES OF TRANSPORTATION.

The ATLANTIC, MISSISSIPPI AND OHIO RAILROAD, running
through from Norfolk, west and southwest, to all points south
and west, passes through the heart of the county. This
great line is of ample carrying capacity, being of five-feet
gauge, and, once out of its financial embarrassments, may do
much more than formerly to develop the material resources
of the section through which it runs.

There are proposed lines of transportation, organized for
the purpose of developing the great ore belts; prominent

among which are the southern extension of the New River
Railroad, the Virginia and Statesville Railroad, and the road
of the Lobdell Car Wheel Company.

FISH CULTURE.

Since the adoption by the Fish Commissioner of a site near
Wytheville for a hatchery, fish culture is assuming some importance.   At a fine large spring, of proper temperature, near
the Atlantic, Mississippi and Ohio Railroad, three and a half
miles west of Wytheville, all the necessary buildings for this
interesting business were completed in 1879, and the young
fish of many improved varieties have been shipped to various
streams in this and other parts of the State.   The whole
matter has been a success, under the intelligent management
of PROFESSOR McDONALD, and now that the hatchery is in successful operation, only a short time must elapse before all
the streams of the county and the section will be stocked
with varieties suitable to them.

PRODUCTION OF CATTLE, SHEEP, WHEAT, PIG METAL, LEAD, AND ZINC ORE.

Fat cattle, annually shipped	1,800.	
Stock cattle " "	2,000.	
Sheep, " "	5,000.	
Wheat, " "	180,000 bushels.	
Pig metal (chiefly for car-wheels)	8,000 tons of late, which will be increased hereafter.	
Lead from Wythe Lead and Zinc Mines.	1,000 tons annually.	
Zinc ore " " " "	50,000 tons have been mined up to date.	
Zinc ore from the Bertha Mine	2,100 tons to date.	

QUOTATION FROM HOWARD SHRIVER, A.M., OF WYTHEVILLE, ON THE FLORA AND CLIMATE OF WYTHE COUNTY.

"Owing to the altitude of Southwest Virginia, averaging
half a mile above sea level, the climate resembles that of the
Middle States, many of our plants belonging to Pennsyl-

vania and New York.   Whereas, in the same latitude, east of
us, near the seashore, the fig ripens its two crops, and plants
common to North Carolina are found.

"Hence, it is less necessary to enumerate the prevailing
plants, and we shall confine our list to a few that may be regarded characteristic of the Flora.   Among the first and
most attractive, is the splendid Rhododendron Catawbiense, Michena, which abound on the hillsides and often extend to their summits, being sometimes intermixed with R.
maximum L. and Kalmia Latifolia—common farther north.

"Along the streams are Clethra Acuminata and Alnifolia,
Andromeda Floribunda Purch. and Magnolia Fraseri, Walt.,
Umbrella Lam. and Acuminata L.   The mountain roads
are lined in the early spring with the fragrant white flowers
of Leucothœ Recurra Buck.

"Among a large number of Vaccinia are found V. Ery-
throcarpon Michx. and Macrocarpon.   Of the Azaleas, there
are A. arborescens, Purch., Viscosa L., Nudiflora L., and the
most splendid Calendarlacea, Michx. in every conceivable
variety of coloring.   Galax Aphyllea beautifies the woods in
winter, with Chimaphila and several Pyrolas.   In summer
the Monotropa, and sweet-scented Schweinitzia Odorata, Ell.
Ilea is well represented, including I. Monticola, Gray., Hale-
sia Tetraptera is found in the water-courses.   The Labiatœ
are well represented, including the somewhat rare Cedronella
Cardata, Benth., Scutellaria Versicolor, Pelosa, Servosa, etc.
Stachys Cordata, Ridd.   Ararum Canadense, L., Virginicum
L., and Arifolium Michx. are common with Aristolochia Sipho,
L'Her. and Serpentana, L.   Pyrularia Oleifera, Gray., Eu-
phorbiæ of various species, including Commulata, Eng., Unu-
laria Grandiflora, and Lessifolia, L. are common, as also
Prosartes Lanuginosa, Don. and Clintoria Borealis, Raf., Um-
bellata Torr., Convellata Majalis prevails on the mountains,
even to their summits.   Of the Carices, some twenty-five

or thirty are common, while Carea Fraseriana, Linn's is rare.

"The ferns are quite numerous, and, at times, grow in great luxuriance. The whole ground is often covered with a dense growth of Cystopteris Bulbifera, Bernh., the fronds from a foot to a foot and a half in height. In the same abundance are found Aspidium Goldianum, Hook. and Clintonianum; while the open woods are carpeted with A. Pedatum Phegopteris, and A. Marginale, Sw., and Acrostichoide, Sw., with occasionally Adiantum Capillus Veneris. The rock crevices exhibit Asplenium Montanum, Willd., A. Ruta Muraria L., Cheilanthes Vertita, Sw., Perlaca Atropurpurea, Link., and Polypodium Vulgaro and Incarium, Sw., A. Tricomanes L., Camptosorus Rhozophyllus, Link. In all some thirty species."

MONTHLY AVERAGE TEMPERATURE AND GENERAL AVERAGES FOR MONTH AND YEAR.

BY HOWARD SHRIVER, A.M., WYTHEVILLE, VIRGINIA.

	Jan.	Feb.	Mh.	Ap.	My.	Ju.	Jy.	Au.	Sep.	Oct.	No.	Dec.	Av. Yr.
1860	37	41	43	52	65	68	65	62	55	53	33	33	54
1861	41	40	41	51	60	62	73	71	65	55	43	40	54
1862	33	31	45	56	61	64	73	74	67	56	44	35	55
1868		31	31	50	50	65	74	69	64	53	40	33	53
1869	35	35	45	51	57	68	74	74	64	46	36	34	50
1870	35	37	40	52	56	68	73	73	64	53	41	31	53
1871	36	40	49	55	62	71	73	73	64	55	37	31	54
1872	30	37	31	48	64	69	73	71	61	56	40	30	54
1873	33	39	39	52	61	70	73	73	64	49	39	39	54
1874	37	39	44	49	63	68	73	68	61	49	41	41	52
1875	34	31	41	50	62	68	75	72	48	49	43	26	53
1876	41	40	41	50	62	68	73	70	62	51	43	41	53
1877	31	39	37	52	56	64	74	71	66	55	43	24	53
1878	31	39	49	55	66	64	74	70	62	51	42	42	53
1879	31	31	43	50	62	65	72	76	60	59	44		53

| 36 | 37 | 43 | 52 | 60 | 68 | 73 | 71 | 63 | 53 | 41 | 35 |

RAIN-FALL, AND AVERAGES FOR MONTH AND YEAR.

BY HOWARD SHRIVER, A.M., WYTHEVILLE, WYTHE CO., VA.

	Jan.	Feb.	Mch.	April	May	June	July	Aug.	Sept.	Oct.	Nov.	Dec.	Avge.
1860	3.1	3.6	3.3	3.6	3.4	3.1	4.1	3.7	2.6	3.4	4.3	4.4	47.0
1861	6.3	5.2	2.3	5.6	3.6	3.2	8.1	3.1	3.3	5.8	4.3	0.2	51.1
1862	4.6	8.0	4.0	6.3	3.2	5.9	4.6	3.2	0.3	2.8	0.5	1.1	41.7
1868	3.1	1.7	2.3	3.1	2.8	2.5	6.6	3.1	4.3	2.6	1.3	3.2	38.2
1869	2.2	3.8	3.2	1.5	4.1	5.0	3.2	2.6	4.2	2.7	1.6	4.1	34.4
1870	2.8	3.8	4.3	4.3	4.7	5.5	4.5	2.6	1.3	2.1	1.6	1.4	37.6
1871	2.3	3.1	4.0	4.3	4.9	3.2	1.7	3.5	3.3	1.7	1.5	6.0	38.1
1872	1.3	5.4	3.0	4.8	5.7	4.9	2.3	3.5	3.5	1.2	2.5	2.9	34.9
1873	5.7	4.7	3.7	2.5	2.3	3.8	5.2	3.7	3.5	1.1	2.2	2.3	44.5
1874	3.7	3.1	6.5	4.5	1.1	9.1	7.8	4.6	2.4	0.5	3.8	1.2	39.0
1875	3.1	4.0	3.0	0.5	3.4	4.0	4.1	2.6	1.5	1.9	3.4	2.7	49.0
1876	2.1	3.0	2.6	5.4	3.0	4.0	4.1	1.9	3.3	3.0	5.4	1.0	33.1
1877	4.0	0.3	2.5	4.4	5.0	4.0	6.4	3.8	2.8	2.3	2.1	4.1	46.8
1878	5.7	2.7	3.2	4.4	5.0	2.0	2.7	4.5	2.8	2.2	3.8	4.2	52.5
1879	4.1	2.1	1.3	2.1	5.1	2.0	2.7	4.5	1.8	3.2	1.1	5.4	38.9

| 3.4 | 3.4 | 3.6 | 3.9 | 3.7 | 3.8 | 4.5 | 3.7 | 3.3 | 3.0 | 2.9 | 2.9 | 41.9 |

MAXIMA AND MINIMA OF THERMOMETER.

BY HOWARD SHRIVER, A.M., WYTHEVILLE, VA.

Year	Jan.	Febr.	March	April	May	June	July	August	Sept	Oct.	Nov.	Dec.

Average Temperature for Month for a period of thirteen to fourteen Years, Observations taken at or near 7 A.M.,

2 P.M. and 9 P.M., until Oct. 1st, '78, subsequently

by Self-registering Max. and Min. Instruments.

RANGE OF BAROMETER.

Reduced to 32° F. and corrected for Capillarity, Temperature, and Variation in level of mercury in cistern. Instrument (No. 1560 by J. GREEN, N. Y.; except from Nov. 1, '65 to Apl. 1, '69; and Sept. 1, '69, to Sept. 1, '72, during which time an ordinary instrument was used). Observations previous to Nov. 1, 1876, are reduced by the constant—.145 to correspond to current observations at present station.   Add 27. to the thousandths in Table.

BY HOWARD SHRIVER, A.M., WYTHEVILLE, WYTHE CO., VA.

	Jan.	Feb	Mch.	April	May	June	July	Aug.	Sept.	Oct.	Nov.	Dec.	Avg.
1865......												569	588
1866......649	649	921	579	512	597	648							
1870......631	473	521	574	571	608	659	655	697	686	635	591	608	
1872 597	518	532	641	631	639	666	669	646	624	597	632	618	
1869......537	490	567	570	470	634	647	694	749	591	582	675	600	
1870......631	473	520	574	571	608	659	654	697	686	635	691	617	
1871......711	604	580	539	604	604	638	630	717	726	596	646	633	
1872......548	504	553	478	545	619	693	666	676	628	512	671	591	
1873......664	620	568	552	573	627	665	632	678	680	692	676	655	
1875......536	568	553	511	569	645	645	615	652	602	590	526	584	
1876......687	618	522	572	626	605	685	675	576	575	612	526	609	
1877......631	610	504	495	500	610	620	600	639	624	600	688	593	
1878......540	459	521	403	541	568	627	570	700	633	546	518	552	
1879......579	547	600	502	624	616	622	598	697	742	678	635	620	

MONTHLY AVERAGES.

Jan. Feb. March April May June July Aug. Sept. Oct. Nov. Dec. Avg.
612 549   548   541 559 611 654 642 675 641 600 619 604

FURNACES AND FORGES IN WYTHE COUNTY, VIRGINIA—ALL COLD-BLAST CHARCOAL.

Cedar Run Furnace, Graham & Robinson, near Graham's Forge (Graham's old furnace), built by David Graham. One stack 32x9. 1832. Water power. Capacity 6 tons.

Barren Springs Furnace, J. Williamson M'Gavock, near Carter's Ferry, New River (Graham's new furnace), built by David Graham in 1853 and rebuilt by Graham & Robinson, 1873. One stack 35x8, cold blast. Capacity 5 tons.

Eagle Furnace, or Gray Eagle Furnace, built in 1863 by Buford, Stuart & Co., now owned by Graham & Robinson.   One stack 33x9.   Cold blast, water power.   Capacity 5 tons.

Brown Hill Furnace, built by Abraham Painter & Sons, 1870, now owned by the Lobdell Car Wheel Company, Wilmington, Delaware.   One stack 32x9.   Cold blast, steam power.   Capacity 8 tons.

Walton Furnace, built by Howard & Saunders, 1872, now owned by Lobdell Car Wheel Company, Wilmington, Delaware.   One stack 33x9.   Capacity 8 tons.

Ravenscliffe Furnace, Crockett & Co., one old stack 29x9, built in 1840, rebuilt 1876, and a new stack 33x9.   Water power.   Capacity 14 tons.

Speedwell Furnace, D. E. James & Son, built 1873.   One stack 32x9.   Water power.   Capacity 6 tons.

Wythe Furnace, Sayers, Oglesby & Co., built in 1873.   One stack 33x9.   Steam power.   Capacity 5 tons.

Irondale Furnace, Noble, Allen & Co., built 1880-81.   P. O., Crockett's Depot.   Capacity 10 tons.

Beverly Furnace, Crockett & Co., built in 1880.   Water power.   P. O., Crockett's Depot, 36x10.   Capacity 12 tons.

Furnace of the New River Iron Co., at Pierce's Falls, New River, built in 1881.   P. O., Jackson's Ferry, 34x10.   Capacity 12 tons.

Furnace of the Hendricks' Bros., of New York, 1881.   New River Mineral Co., now building at mouth of Painters' Branch, New River, two miles above Wythe Lead Mines.   One stack 10½x40.   Capacity 20 tons.   P. O., Brown Hill.

Forges.—Of the numerous forges once existing only one remains, Graham's Forge Rolling Mills and Nail Works, built by David Graham, on Reed Creek, 1828.   Three heating furnaces, four trains of rolls, five nail machines, and one hammer.

SMYTH COUNTY.

Smyth County, for the princely resources it contains, has been more neglected comparatively, hitherto, in all reports having the sanction of authority, than any other county in the State. The great Valley of Virginia, as widely renowned as it is for the beauty and fertility of its soil, and its matchless wealth in ores, would be incomplete indeed without the county of Smyth. The citizens of the southwestern end of the valley, being very generally acquainted with the character of what may strictly be termed the Shenandoah end of the valley, live also in the knowledge of the extraordinary resources in Salt and Plaster, Lead and Zinc, and great masses of Magnetic and Brown Iron Ores, which add such an additional interest to their own end. Could there be that capable and trustworthy management of public property that there is ordinarily of property in private hands, what a source of revenue would there be to the county, as an owner, or the State, of the vast salt interests at Saltville, and the great plaster beds there and above, on the Holston River and tributaries. How many of the burdens would be thus removed from the people of a county, with a revenue coming into its treasury, annually, equal to its receipts from taxation. Such a revenue would not only do away with the taxation, but a great part of it would be spent among the people for the building of many needed improvements. Though it is now too late to think of such a possibility for the salt works, could the great plaster beds even, on Cove Creek, become the property of the county, and be leased out to competent and honest men to be mined for its benefit, cheap railway transportation being understood, what a source of revenue would thus be opened.

If this railway transportation would open up Kentucky and

its connecting lines of railway, it would be difficult to estimate the vast quantity of this cheap and abundant fertilizer
which would be annually consumed. But, pleasant as it
might be for the county to have an independent source of
revenue that would free it from the necessity of taxing its
people, would it not be an impracticable thing, because of
the impossibility of employing agents to handle the property
who would, with single-hearted fidelity, conduct the management as judiciously as they would in the position of private
owners? Could the county, under the law, become an owner,
the results of the venture might be anticipated as of rather
doubtful success, unless the properties could be handled with
address and fidelity.

However this may be, Smyth County, not merely because
of the existence of these remarkable deposits of plaster and
salt, is great. There are vast deposits of iron and manganese
of superior character; and these magnificent veins and deposits lie sometimes within, and generally but a mile or so
from grass and grain lands unsurpassed in fertility.

Could the resources of Smyth County be utilized, so really
abundant are they that it is not too much to say the finances
of the State itself would feel the beneficial effects of such
development in a marked degree.

This is another one of those counties to which it will be
impossible to do justice in anything short of a volume. A
full description of the Plaster and Salt deposits alone would
require much space; but in the section allotted to the
county, enough, it is hoped, may be said to draw attention to
its highly valuable resources; and that, too, if the design
should be successful, in a manner which will show their
bearing upon the general prosperity of the country.

HOW BOUNDED.

Smyth is separated from Tazewell, on the north, by the Clinch range of mountains; on the south, from Grayson, by the Iron Mountain, the southwest corner being on the great White Top; on the east side it is bounded by Wythe, and west by Washington.

HOW WATERED.

The county is watered almost wholly by the different branches of Holston River, two of which—the Middle and South Forks—take their source in the county; but in the head of Rye Valley are some of the head-waters of Cripple Creek, which creek flows eastwardly to New River, while the Holston waters flow southwestwardly toward Tennessee.

GEOLOGICAL.

The geology of Smyth is comprised between the Upper Huronian Rocks, showing in White Top Mountain, and the Proto-Carboniferous, showing in Brushy Mountain; in one place there being an exception—at Saltville, where there are evidences of the Cenozoic, or Mammalian Age.

With the exception of the Salt and Plaster, there is no marked difference between this county and other counties of the Great Valley, either in its geology or mineralogy. The cross section, here given, will show the geology of the Valley of Virginia somewhat differently disposed from what it is at other points east, but not essentially different in character. A description of the geological section here given is scarcely necessary, as it explains itself with sufficient clearness for general purposes. The county, really, would require a number of cross-sections to show the dip of the rocks in each district; but the section given will show the general order

and position of the various strata, faults, etc. It may be
proper, however, to call attention to the series of mountains
near the middle and in the eastern part of the county : three
high ridges, running side by side, a course north 70° east,
dying down near Marion, at the west end, and breaking off at
the eastern end near the county line, forming a great island
of Potsdam sandstone, iron ore, and red shale, surrounded
by the valley limestones. Likewise Big Walker's Mountain,
which is such a vast barrier in the eastern end of the county,
as it approaches Saltville becomes a mere hill, and after that,
going southwest, has no longer the same character. The
other mountain ranges are continuous, except White Top,
which is a massive and towering outburst of granite rock
material, standing almost isolated.

IRON ORES.

It would seem as though we were to have a redundancy of
the subject of iron ores in this work. In each county it has
been a prominent feature! In Smyth the measure will be
found full and running over, and the only regret will be that
of the writer at his inability to present the subject as it deserves.

Beginning on the south boundary of the county and going
northwardly, the first great extensive line of iron ore deposits
are the brown ores at the base of the Calciferous and the upper part of the Potsdam sub-epochs in the spurs and parallel
ridges of the north side of Iron Mountain. These ores combined with manganese ores form heavy beds and deposits in
this range, apparently more profuse as to surface indications
toward the eastern than the western end. That the massive parent veins from which these decomposed ores are derived are some of them over 50 feet thick, I have no doubt.
Pearce's Forge, near by, on the South Fork of Holston River,

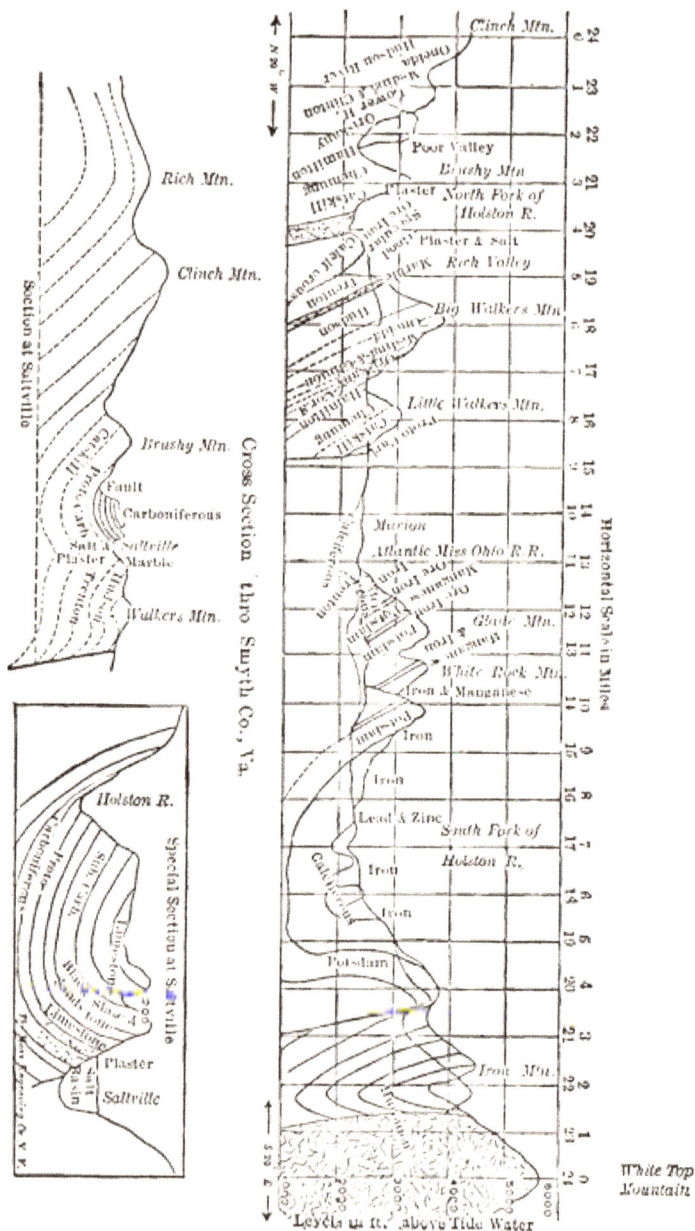

Section at Saltville

Rich Mtn.

Clinch Mtn.

Brushy Mtn.

Fault
Carboniferous
Salt & Saltville
Plaster Marble

Walters Mtn.

Cross Section thro Smyth Co., Va.

Special Section at Saltville

Holston R.

Limestone
Black Slate &
Sand-stone &
Limestone
Plaster
Saltville

Clinch Mtn.

Poor Valley
Brushy Mtn.
Plaster North Fork of
Holston R.
Plaster & Salt
Rich Valley

Big Walkers Mtn.

Little Walkers Mtn.

Marion
Atlantic Miss Ohio R R.
Iron
Iron & Manganese
Glade Mtn.
White Rock Mtn.
Iron & Manganese
Iron

Iron

Lead & Zinc South Fork of
Holston R.
Iron

Iron

Potsdam

Iron Mtn.

Horizontal Scale in Miles

Levels in ft. above Tide Water

White Top
Mountain

derives some of its ores from either the parent or derivative beds of this line of ores. Much of this ore is no doubt
the result of the decomposition of the heavy measures of sulphides of iron, lead and zinc, which once extended in masses
up and down this valley, in a prolongation southwest of similar great deposits in Wythe ; and although developments there
are still in the very earliest stages of the initiative, enough may
be seen, by a careful search, to prove that there must still remain somewhat of this same formation, here and there, sufficient to warrant the belief of paying quantities of zinc still
remaining, particularly toward the middle and east end of
Rye Valley.

Then, crossing over to the north side of the South Fork of
Holston, along the strike of the northern out-crop of the great
synclinal—of which the river and Rye Valley are the marks of
the greater depression—we see another long line of beds of
brown iron ore, parallel generally with the course of the river
and valley, and showing in many places profusely from the
Smyth-Wythe line to Thomas's on the Marion-Grayson turnpike. These ores are likewise here and there mixed with
manganese, but often a pure brown hematite, so-called, of a
high degree of purity and excellence, and in quantities corresponding to the vast parent veins from which the ore was
derived. These ores are frequently pocketed in the limestone, and when so found are regarded as the purer ores.

Next of importance are the almost inconceivably vast beds
and deposits in the series of mountains running from the
immediate neighborhood of Marion, eastwardly, to the Smyth-
Wythe line, in which the Glade Mountain, with its extensive
veins, bears so important a part. Up Staley's Mill Creek,
Nick's Creek, heads of Aker's Creek, Phillippi's and Steffee's
branches, in Glade Mountain ; in the White Rock Mountain
and the space between it and the flanking ridge on the south,
and in that district, where these mountains die down under

the limestones just south of Marion, the deposits and veins of iron ore, from very careful and repeated observations, are judged to be among the first in the world in size, in facility of mining, and easy accessibility. This is saying no more than the facts warrant ; and although it will be impossible to give, in this space, the exact measures of all the beds—those fine ones nearer Marion, and those nearer the White Rock Furnace (formerly the Panic Furnace) the measures of just a moiety it is hoped will be convincing. Taking from the field notes we have : "Head of Steffee's Branch," a heavy body of iron ore following the strike of the rocks a great distance, 130 feet across the vein which runs north 75° east about, on the same lead which gives the fine specular ore, chalcedony, etc. ; barometer 3,200 above sea level ; dip of rocks south 15° east ; stratification as follows : Beginning south a heavy band of sandstone, then specular ore eight inches, then thin bed of sand-rock, then (manganiferous iron ore perhaps a third of the vein, followed by pure brown iron ore) the 130 feet of ore. Then at about 700 yards north, down the mountain (north 15° west) passed another iron ore lead (parallel to the first) 24 feet thick at the division between the Scolithus-marked sandstone and the red shales, etc., of the Lower Calciferous ; then at 500 yards more on the cross section (course north 15° west), struck 10 feet of kidney ore overlying a heavy ledge of sandstone, dip 20°, south 15° east, barometer 2,760 above sea level, which point was 250 feet above the ATLANTIC, MISSISSIPPI AND OHIO RAILROAD, two miles farther north.

Then, again, in the examination up Nick's Creek, the largest bed of good ore, situated at the western end of White Rock Mountain, where Nick's Creek cuts the mountain off, from the surface indications, is judged to be 300 feet thick, dipping southwardly. This vein or deposit, like the others just mentioned, is continuous for miles through the country—for ten miles, if not more—in easily accessible ground. Of good ore

7

in these veins, which might be mined or stripped above ordinary water level in the creeks, there are not less than
25,000,000 tons. Some experts, accustomed to the examination of ore deposits, give the beds nearer Marion alone credit
for more than that amount. The curious and critical may
take the elevations as given by the barometer above, and
only one half the recorded thicknesses of deposits by a length
of ten miles, and readily see whether the quantity is an exag-
geration or not.

Again, as you approach the Atlantic, Mississippi and Ohio
Railroad in the middle of the county, on either side of the
valley of the Middle Fork of Holston River, there are deposits of brown ores which have evidently resulted from the
decomposition of the ore beds at higher levels, and subsequent precipitation in the cavities of the limestones below.

Then going northwardly over some lines of lesser veins in
the flanks of Little Walker's Mountain, the next notable deposits of brown iron ores are along the outcroppings of the
Oriskany measures in the south flank of Big Walker's Mountain. This line of ores, showing generally not far above the
base of the mountain on the southern side, lies in a line fully
twenty miles long, coursing north 60° or 70° east. These
deposits are similar to other Oriskany deposits described in
different parts of this book. Now and then the lead shows
but little ore, and again the surface is covered with an excellent brown ore which has not yet been found to yield on
analysis over three tenths of one per cent. of phosphorus.

Then, again, at Ward's, near Chatham Hill, are brown ores
formed from decomposed pyrites in respectable masses, but
no doubt belonging to the division between the Black River
and Trenton series.

These same conditions, as to Oriskany ores, are repeated
again in Smyth County, in the south flank of Clinch Mountain, and in Poor Valley Ridge.

In truth, the brown iron ores of Smyth, above water level, are not far from 100,000,000 of tons.

Red Iron Ore.

Beginning at the southern boundary line, and going from south to north over the various strata, not much pure red iron ore is met with, until you reach the extraordinary vein in Glade Mountain, mentioned above. This ore—a pure specular, crystalline in structure, very dense and free from impurities—is only, so far as developed, about eight inches thick, with an evident tendency to thicken eastwardly and to thin out in the opposite direction. It is singular that, at the head of Steffee's Branch, a vein of this nature should occur in a regular and well-defined system of rocks, and be confined to only a few hundred feet either way.

The next notable line of red ores met with, going north-wardly, are those occupying the line of felspars at the junction of the Black River and Trenton limestones. This remarkable band of ore, which has such a large development in Giles County, about Chapman's, shows best in Smyth at Tilson's, east end of the county, in Rich Valley. The ore is of the finest quality, and is apparently not less than six feet thick with the rocks dipping southwardly. In nearly every hill, as you go down the valley, this remarkable ore shows itself with the felspar overlying it. It will prove of very great importance to the iron interests of that section, as it is not only a remarkably pure ore, but in great quantity in the aggregate.

After this no very considerable masses of red ores are again met with until you reach Big Walker's Mountain, where the fossil red ores show, some distance below the crest of the mountain, on the south side. In addition to the strictly fossil ore there is a red hematite in the same series

of rocks, slaty in structure, and apparently a valuable ore,
being in quantity now and then. The fossil ore, which dips
southwardly between ledges of red sandstone, rarely ever exceeds eighteen inches in thickness. There is also a short
line of these ores in the south face of Clinch Mountain, and
again in Poor Valley Ridge.

Magnetic Iron Ore.

There is a line of magnetic ores leading from near Marion
southwestwardly toward the Smyth-Washington line near a
point where the South Fork of Holston River leaves the
county. These ores show in fragments on the surface, but
sufficient developments have not yet been made to prove
quantities. The measures seem to be confined to the Lower
Silurian limestones.

Iron Pyrites.

Iron pyrites is, perhaps, the original material of all the
iron ore deposits of the county except the specular and fossil
ores. On the South Fork of Holston River, in Rye Valley,
Glade Mountain, etc., and in Rich Valley, iron pyrites must
be abundant below water level. Cubical pyrites are largely
disseminated in the slates, etc., of the Proto-Carboniferous
series, just north of the fault in which the plaster and salt are
found, as well as in larger masses in the lower limestones
south of the same fault.

COPPER.

Carbonate of copper is distinguishable in the line of rocks
in which the barytes occur, just south of the Atlantic, Mississippi and Ohio Railroad, four miles east of Marion. Copper pyrites are, no doubt, in the same series. In fact, following the same series eastwardly nearly to the Smyth-

Wythe line, copper carbonate and sulphide are so abundant as to create the impression of large quantities below the surface. Copper pyrites ought also to be found in the line of rocks in Rye Valley and south fork of Holston River, which represent the lead and zinc-bearing formation.

Again, in the Hamilton slates, in Lick Creek Gap, in the northern portion of the county, and at points in the same strata in Poor Valley, indications have, now and then, been found to lead to the belief that sufficient quantities existed to pay for mining, but this is doubtful.

LEAD AND ZINC.

It is unfortunate that the Lead and Zinc indications showing at some points near Sugar Grove, in the South Fork Valley, should not have been prospected to such an extent as to permit measures to be taken. Without more evidence, the presumption is that there is here a continuation southwestwardly of the heavy lead and zinc-bearing strata showing so conspicuously in Wythe County, the general geological formation being identical. Lead has been occasionally found in Rich Valley, near the line of the division between the Calciferous and Trenton rocks. Again in the south of Big Walker's Mountain, in the Corniferous flint on Bear Creek, and in the Lower Helderberg group. The same may be said of the like formations in the south flank of Clinch Mountain. These latter ores have often been claimed to be silver-bearing galenites, but Dr. Genth's analysis failed to find the silver.

SALT AND GYPSUM, OR PLASTER.

In treating this highly valuable and important subject, it will not be inferred that the author regards his statements concerning either the geology or other important features

of the formation infallible; but believes, from the fact that
his examinations have continued over a longer period than
most others have been able to devote to it, that his conclusions are safe.

Saltville, the place where salt has been manufactured,
without intermission of any duration, for a great many years,
including the ground occupied by the Buena Vista Plaster
Mills, is the southwestern limit of the extraordinary deposits
of salt and plaster which mark the line of a great fissure in
the crust of the earth, along the course of the North Fork of
Holston River—mainly in Smyth County—about seventeen
miles of which show Salt and Plaster; but only that singularly beautiful basin surrounding Saltville is positively
known to yield much salt.

This great fissure, it may be as well to say, brings up the
limestones of the Lower Silurian division, charged with sulphide of iron, against a downthrow of Proto-Carboniferous
rocks, charged also more or less with sulphide of iron. It is
possible that after the great pressure (from southeast to
northwest) upon the arched crust of the earth had resulted
in the above action, a compensating pressure was then exerted at right angles to the first, causing in many places
great fragments of the crust to slide between each other; in
other places (as the one now treated) causing the great fis-
sures between great fragments to yawn (at the same time the
whole being raised) and remain open, probably because great
pieces were precipitated into the chasm, preventing the sides
from settling back into close contact with each other. Into
such a great yawning series of chasms, on the line above
mentioned, since the close of the Carboniferous age (when
the fracture is supposed to have occurred) have the waters
coming from the surrounding strata, charged with brine from
the salt series of the coal rocks, and with sulphuric acid and
lime from the decomposing iron sulphides and limestones,

been pouring thousands upon thousands of tons of both salt
and sulphuric acid and lime, now combined into sulphate of
lime, or gypsum. These solutions finally deposited the rock
salt and plaster; the salt seeming to have been deposited
first, as at Saltville, over 175 feet thick, its top being at a
depth of 200 feet from the surface. Then above this comes
about 100 feet of bluish slate, overlaid with gypseous clays of
variable thickness. These measures are only local to Saltville. At the Buena Vista plaster beds, a mile or more southwest of the Saltville beds, the plaster is a fine compact body
of great but scarcely known thickness, though it is sometimes asserted that its depth is determined. Northeastwardly from Saltville six miles, immediately on the river, the
Pearson plaster beds have been explored to a depth of about
180 feet; and they indicate not only continuity, but great
solidity. At Buchanan's Plaster Cove, sixteen miles northeastwardly from Saltville, on the north side of Holston River,
the great chasm must have been very wide and deep. Here,
an 8 by 10 shaft, which was sunk ??? feet in search of salt,
was in fine plaster all the way down, showing saline satura-
tion to some extent near the bottom.

While it is not unreasonable to suppose that at Saltville
and the immediate vicinity there are 500 acres of land underlaid with rock salt, there can be no impropriety in giving the
quantity of it underlying 100 acres. Such a calculation will
serve to show why these deposits have been drawn upon so
long without apparently losing anything of their original
strength and quantity.

The most reliable data give the thickness of the rock-
salt at 175 feet, with the certainty of much of the overlying
and underlying material so heavily saturated with brine as
to almost, if not quite, form a source of supply equal to one
half the volume of the rock salt; but, discarding this view of
the case as not fully proven, the calculation upon the basis

of 175 feet thickness will make manifest the inexhaustible source from which the Holston Salt and Plaster Company are deriving their salt. For 100 acres the result is about 2,100,000 tons of rock salt. It is true, that in the 175 feet there will be a considerable quantity of earthy as well as rock material; not only is this saturated with brine, but enough of the surrounding material to justify the calculation being made upon the basis of 175 feet thickness, solid. Now it may be assumed, without reasonable doubt, that there is at least five times that quantity, with the very high probability that the rocks of Nos. X. and XII., the original source of supply, are still by drainage annually adding fresh supplies.

It may be submitted, then, that an annual consumption of 23,000 tons of salt, as is the present yield, will not exhaust the supply under 100 acres of land for seventy years to come, assuming that in the past the quantity consumed is about equal to 500,000 bushels annually for thirty years; and if the rocks of X. and XII. are still giving up salt, it is truly inexhaustible. It has never been necessary to pump in fresh water, as sometimes asserted, in order to keep up a regular flow through the pumps. The water, finding its way to the salt alone as fast as needed, soon takes up the necessary quantity, and comes out saturated to the usual density, which is now given at 98 per cent.

GYPSUM.

As to the quantity of the gypsum, if it were solid, 80 feet thick, it would yield about 90,000 tons to the acre. But there is really no telling how much ground about Saltville is underlaid with plaster. It may be confined to the edges of this basin, or, if deposited from a solution, as is strongly suspected, it is likely to underlie the whole acreage of the Salt-

ville basin. In this case the quantity is far beyond any demand which is ever possible to be made upon it. At the
Pearson plaster beds these conditions are likely to prove the
same. At Buchanan's Plaster Cove, sixteen miles east of Saltville, the plaster is known to be 592 feet thick at one point; and,
in all probability, underlies an acreage fully as large, if not
larger, than that at Saltville. One acre of it, to the depth
above given, from actual measurement, holds over 666,000
tons. One hundred acres will yield 66,600,000 tons, if the
data derived from close investigation be admissible.

The acreage under which this extensive gypsum deposit
has been positively ascertained—in the Saltville Basin,
about Pearson's, Taylor's, and the Buchanan Plaster Cove
—may almost be estimated by the square mile rather than
by the acre. To give the acreage would therefore be superfluous.

The present annual consumption of plaster, from all these
deposits, is about as follows :

Holston Salt and Plaster Co............ ....3,000 tons.

Buena Vista Plaster Co.............. 2,000 "

Pearson Plaster Banks .... 800 "

Buchanan or Cove Plaster Banks........ 500 "

Total.......................6,300 tons.

The first two plaster companies enjoy railroad facilities,
supplied by the Atlantic, Mississippi and Ohio Railroad Company; the latter two hope that within a year or two, either
the RICHMOND AND SOUTHWESTERN RAILROAD, or the VIRGINIA,
KENTUCKY AND OHIO RAILROAD will be built through both their
properties. It may not be inappropriate to say, before dis-

missing the subject of plaster, that its use as a fertilizer, in
Virginia alone, should be much greater than it is; not only
because its consumption would save annually to Virginia a
very large sum, which is now being sent out of the State for
other fertilizers not as good; but for the reason that, the
State finances being low, so fine a basis upon which to create
much more extensive operations, and much more the ability
to pay taxes than now exist, should not be neglected.

Now, it is plain that, without other lines of transportation
even than those at present in use, this desirable object can
be to some extent accomplished, and that will be by causing
the plaster to be used as a permanent fertilizer, instead of a
mere stimulant for growing crops, as is now generally the
case. The following being the reasons for the assumption
held forth above:

Plaster, or the ordinary gypsum, contains about forty-five
per cent. of sulphuric acid. When finely pulverized and inti-
mately mixed with any soil it is clear that this acid must be
a solvent for many of its constituents, all, in fact, upon which
it will act anywhere. Now the soils of this country contain a
good deal of silica combined with potash and soda, as well as
iron ores in greater or less quantity, more or less charged
with phosphorus. It is plain then that if a cheap reagent
can be had which will dissolve these substances, and cause
them to yield the potash, soda, phosphorus, etc., we have all
the requisites. It is here submitted that gypsum will do
this, if it is a good soluble article. The reactions are likely
to be something like the following: The gypsum thrown into
the soil in a pulverized condition, and coming into contact
with moist substances, for which its constituents have a high
affinity, it begins to break up or dissolve, the sulphuric acid
attacking mineral and organic matter alike, together with the
downfall of ammonia in rains and snows, causing it to take
forms highly useful to plant life, which the plant couldn't utilize

before. Thus the flint gives way and yields potash, soda, and silica ; clays are dissolved to some extent and form new combinations ; the iron ores give up their phosphates, as well as the minute fragments of apatite; and the lime, left to itself as pure calcium, forms other reagents, prominent among which is calcium hydrate or caustic lime, which is itself one of the most powerful reagents known.

All of these acting together do all the work nearly, in the soils of Southwestern Virginia and the neighboring States, which any fertilizer can accomplish. Let us see then if there is any prominent practical illustration of the theory.

Mr. Legrand Sexton, of Chatham Hill, in Smyth County, having bought two old worn-out farms near the Cove Plaster Banks, determined to try the efficacy of plaster as a permanent fertilizer, since he was so close to the deposits as to bring the question of cost to a nominal figure. Upon these two places the soil had been reduced so by constant cropping that they would scarcely produce nine bushels of corn to the acre ; the soil being in the limestone belt, and covered more or less with flint containing potash and soda and doubtless much comminuted iron ore.

At first he plowed four inches deep, sowing about one bushel of plaster to the acre ; the next year he gauged his plows two inches deeper, sowing an increased quantity of plaster, the yield of corn—the crop he used—greatly augmenting ; the third year he plowed two inches deeper, about eight inches altogether, using something less than two bushels of plaster to the acre, his crop of corn at the end of the season being so great as to astonish him. The fourth year he plowed still deeper, bringing up the clay subsoil into contact with an increased quantity per acre of plaster, making a yield, at the end of the fourth season, of one hundred and twenty-five bushels of corn per acre, on ground that had been really abandoned by the unenterprising people who had previously held it. This land when last seen seemed to be in a state of permanent fertility, for the corn on it in 1878 looked to the writer as though it would yield over one hundred bushels to the acre.

It may be inferred from these facts that, should the owners of plaster deposits who can afford it send out competent agents to all points accessible by rail, and there lecture upon the true character and capacities of plaster, demonstrated both by chemistry and practice, they would thereby so greatly increase the demand for a good article as to make it difficult to supply it. As long, however, as the active canvassers of fertilizers from enterprising Eastern firms go through the country, making statements sometimes prejudicial to the use of plaster, and altogether in praise of their own wares, without meeting men of like activity and enterprise working for the plaster interest, we may conclude that just so long will the important industry of plaster development remain at a standstill, if it does not recede.

BARYTES.

This has been another source of revenue to the county of Smyth. An enterprising gentleman from New York State, some years since, finding barytes in large quantities near Marion, began its mining and shipment. From this business he has made large profits ; and, by his work, has demonstrated that the county, particularly in the localities just east and west of Marion, is capable of yielding a very large tonnage of this material. It seems to be pocketed in Lower Silurian limestones, in a series parallel with and just south of the Atlantic, Mississippi and Ohio Railroad. It is sometimes picked up in other sections of the county, but, as yet, not in sufficient masses to justify the expectation of large quantities.

MARBLE.

Smyth holds two lines of rock yielding handsome specimens of variegated marble. One is a gray variety, closely
resembling that from Tennessee, found in a railroad cut one
mile east of Marion; the other belongs to the series of rocks
near to and just south of Saltville, and all the way up Rich
Valley, which lie about the junction of the Trenton and
Hudson sub-epochs. In these ledges may be discovered
marble of purple and brown variegation, and a gray variety
also. Both of these would polish well, apparently, and yield
a handsome ornamental stone. Should they prove sufficiently
firm in large blocks, there is scarcely any limit to the
quantity.

KAOLIN.

The mountainous region between the Atlantic, Mississippi
and Ohio Railroad and Rye Valley shows beds of kaolin of
unknown extent. The quality, also, has not yet been tested.
It is supposed to result from the decomposition of strata
largely composed of felspar in the Potsdam rocks.

TIMBER.

This county can enumerate all the valuable kinds of timber
known to the latitude, including two or more varieties of the
fir tree. White Top Mountain has a large body of balsam
fir; with this may be placed the lashhorn, a kind of fir with
a differently shaped leaf from the ordinary balsam fir, and
having its limbs to grow out from the body of the tree in
such a manner as to form a lace work, apparently lashed
together so thickly as to permit a number of persons to
occupy the flattened top without danger of breaking through.

All over the sides of White Top Mountain are vast forests
of a superior growth, including much wild cherry, poplar,

etc. In the Iron Mountain there are fine bodies of white pine, and all through the county, except right on the line of the Atlantic, Mississippi and Ohio Railroad, and the vicinity of Saltville, there are immense primitive forests miles in extent. Such is the case in the north spurs of Iron Mountain, Glade and White Rock Mountains, Little Walker's, Big Walker's, Brushy, Clinch, and Poor Valley Mountain, and adjacent valleys. So that the iron master need not fear the early disappearance of an abundance of charcoal. In many of the limestone districts there are still remaining very large quantities of walnut, sugar tree, and other fine woods.

WATER POWER.

The three forks of Holston River, each discharging about 150 cubic feet of water before leaving the county, afford sufficient water power with their tributaries to supply all the demand which may ever be made upon them.

These streams are constant.

AGRICULTURE.

The most profitable branch of agriculture is grazing. There are but few large farms employed in raising grain alone. The lands are generally well adapted to grass, bringing blue grass naturally. A very large area of the county is taken up by high mountains; but the area of limestone, or strictly grass lands, is more than half the county. The valleys of the North, Middle, and South Forks of Holston River, including Rich Valley on the north side of Walker's Mountain and Rye Valley on the south side of the county, show all the fine features characteristic of the best lands of the Valley of Virginia. Sinclair's Bottom, on the South Fork, near the Smyth-Washington line, is often quoted as one of the best bodies of land in the State. Every one who has seen the

Francis Palmer. At 2 Years.

11278. *Albert Edward. At 5 Years.*
TWO OF THE FINE HERD AT SALTVILLE, VA.
(P. 111)

Saltville basin can only speak in terms of praise of the great
fertility of its loamy soil, and thus could many places noted
for the excellence of the soil be called over in the county.
Poor Valley makes no pretensions to richness of soil, but
many farms are situated in it notwithstanding, the people
living well. It is superfluous to name the different farm products of the county: wheat, corn, oats, hay, rye, and buckwheat are the ordinary crops throughout, and no season is
remembered when there was a complete failure in any crop.
Cattle, horses, sheep, and hogs do well, some of the finest
herds of cattle in the country being now raised.

SCENERY.

Smyth, Washington, and Grayson counties may all make
equal claim to the incomparable scenery spread out on every
side of the White Top Mountain. To confine one's self to the
magnificent picture presented to the tourist on first seeing
the carpeted summit of White Top, relieved by a background
of fir and lashhorn trees, is itself sufficient to stamp the place
of the first order. This grassy plateau, on the top of the
mountains, of one hundred acres, is covered over with a thick
and deep turf of some grass usual to high altitudes, and unknown in the valleys below ; watered here and there by crystal springs, it affords, in spring and summer, fine, nutritious
grazing to stock, etc., which only serve by their presence to
heighten the effect of the rare, splendid picture of the green
field with its setting of darker green, 5,500 feet above the
sea.

Then, looking through the clear atmosphere on any side,
with the field-glass, or without it, a rare and lovely landscape
meets the eye. In the distance may be seen the pigmy looking railway trains, apparently moving at a snail's pace,
though going at high speed, leaving behind them trails
of smoke of deeper and lighter shades.

The open farms look like blankets of green upon a great surface of darker colors.   Only by seeing it can the view be appreciated.

Saltville stands in broken groups in a basin, cut by the hand of nature out of an emerald.   This lovely vale sits the mistress of all scenery! Beyond the power of description, a vocabulary of praise would hang like an ugly web upon its quiet beauty.   It is sweetest nature in its noblest moment frozen into eternal repose.   It is a poem of Heaven, making music in the hearts of the glad and the sad alike.   It is the last and most beautiful touch of the Almighty Hand, renewed every year in changing hues, speaking plainly, "This is *my* handiwork! Behold it!"

So are the many other beautiful pieces of nature's painting in this county, but all must yield the palm to Saltville Basin.

Not in the days when it was but a salt, salt sea did it give promise of its resurrection, in these latter days, into such a living source of beauty and profit.   What a pity that man should mar it, or make other use of it than for the glory of Him who made it.   Not that we say it is being used otherwise now, for it gives of its substance to thousands.

To attempt to eulogize its beauty would be to multiply words without the power to touch the subject; so we leave it, radiant in its own power, to best proclaim itself unequaled.

MINERAL SPRINGS.

At this time the Chilhowie Holston Springs, on the Middle Fork of Holston River, ten miles southwest of Marion, are the principal springs prepared to keep visitors.   Their waters have not yet been fully analyzed; but their general character, together with the fine air and lovely river scenery, combine to render the place attractive.

MANUFACTURES.

The county is not only supplied with the requisite number of good grist and saw mills, in various neighborhoods, tanneries, etc., but it has two first-class woolen factories, besides the great manufactory for salt at Saltville.

The Holston Woolen Mills, situated about six miles southwest from Marion on the South Fork of Holston River, now annually turn out large quantities of cassimeres, jeans, blankets, etc.

The factory in Rich Valley will do the same, and both have become recognized as safe and successful institutions of their kind. Marion has long been known as having one of the best plow factories in the country, supplying a very large trade.

FURNACES AND FORGES.

Panic Furnace, or what is now called White Rock Furnace, went into blast August 9th, 1875, for some years out of blast, but is again at work under its new ownership, that of the Lobdell Car Wheel Co., of Wilmington, Del., as a cold-blast charcoal six-ton furnace. It derives its ores from the large beds in the slopes of White Rock and Glade Mountains and vicinity; and will perhaps command some of the magnetic ores of Grayson and Ashe counties, it being proposed to connect this furnace, the magnetic ores mentioned, and Rural Retreat on the Atlantic, Mississippi and Ohio Railroad, by means of a narrow-gauge railroad.

Pearce's Forge, three and a half miles below Sugar Grove, on South Fork of Holston River, makes excellent bar iron from the ores of the great iron belt in that vicinity.

This is about the sum of the important manufactories, furnaces, and forges in Smyth now in operation, or contemplated to go to work soon. The old furnace in Staley's Mill

8

Creek neighborhood is scarcely worth mentioning except
from a historical point of view. It seems to have been in
use during the late unpleasantness, and is said to have illustrated the high quality of the ores of that vicinity.

The ATLANTIC, MISSISSIPPI AND OHIO RAILROAD runs through
the county from east to west, or, rather, from northeast to
southwest, bringing the county into communication with the
eastern seaboard, and the western and southern railroads
and rivers. The Saltville branch of this line also leads back
into this county, although it leaves the main stem in Washington County.

In Rich Valley, or the valley of the North Fork of Holston
River, are the projected lines of branches of the Richmond
and Southwestern Railroad, and of the Virginia, Kentucky
and Ohio Railroad. Either of these, if built, will open up
the vast plaster deposits of Buchanan's Cove, and the agricultural and mineral resources of a fine section.

There are also one or two lines of railway chartered to
cross the country, leading both toward the copper mines
south of the Iron Mountain, and toward Tazewell County on
the north side. One of them to the south, the Virginia and
Statesville Railroad, may at some day be built from Adkins,
on the Atlantic, Mississippi and Ohio Railroad, through
Grayson and Ashe, etc., to Statesville, N. C., opening a rich
mineral and timber region. The other, the SALTVILLE AND
COAL MINE RAILROAD, will soon commence construction from
Saltville to the coal measures in Russell, etc.

Fish Culture will finally become a necessary industry in
Smyth County. The streams now have a great many fine
bass, redeye, chub, sucker, and the mountain streams some
trout. Lick Creek, which comes out from the direction of

Burk's Garden, has still some trout, as well as many streams flowing from the Iron Mountain.

The streams are well adapted to game fish, and will at some day be utilized on account of their industrial value in this way.

Bee Culture is carried to some perfection in Smyth, not so much for the profit on the honey raised for market, as for home use. Many improved gums have been tried, of which the Starbuck patent seems to be very generally in use. There is also another one of home invention, known as the Davis hive, which seems to meet the requirements of this latitude.

Grape Culture, which seems now not to attract so much attention, was once a subject of importance. Much wine was made, by the Sprinkles particularly, a few miles east of Marion; but now, except for table use, but little attention is being paid to the improvement of the grape, or to its culture as a wine crop.

TRADE IN CATTLE, SHEEP, WHEAT, CORN, ETC.

Cattle, about .... 2,300 head sold annually—75 thorough-bred annually from Palmer at Saltville.
Sheep, " .. . 3,200 "
Wool, " ....15,000 pounds.
Wheat, " ... 95,000 bushels.
Corn, but a small surplus is sold.
Barytes, 1,600 tons.

EDUCATION.

More than usual attention has been paid to the important question of education in Smyth. There is a fine high school at Marion, the Marion High School for boys and girls. This is evidenced by the numbers that are sent from surrounding counties. Besides this, and other good schools of its kind, the public school system is kept up to as high a state of efficiency as the public funds will permit.

TOWNS AND VILLAGES.

Marion, the county site, is nearly in the center of the county, on the Middle Fork of Holston River, and by it passes the Atlantic, Mississippi and Ohio Railroad. Marion, besides its good school, has timber factories, fine flouring mills, hotels, churches, stores, a bank, and various repair shops, and ought naturally to be a thriving place.

Saltville, as yet a village, is at the present terminus of the saltworks branch of the Atlantic, Mississippi and Ohio Railroad, and is the center of a large trade growing out of the industries of salt manufacturing, plaster mining, and grinding. It has a good hotel, a tasteful church, and handsome residences, store, and numerous salt factories with their appurtenances.

Its trade amounts to about five hundred thousand dollars annually.

When the contemplated system of roads, north and south, shall have been completed by which the great sulphureted beds and veins of Grayson, Carroll, and Ashe counties can be utilized, extensive works for the manufacture of the fine fertilizer, soda ash, will be erected at or near Saltville. England now manufactures many millions of tons of this cheap fertilizer annually.

Seven Mile Ford, on Middle Fork, long known as the western terminus of the macadamized turnpike which leads eastwardly to James River, is now also a station on the Atlantic, Mississippi and Ohio Railroad.

It is a beautiful place, taken with its surroundings.

Adkins, toward the eastern end of the county, and

Greevers, toward the western side of the county, are both thriving places, on the same railroad, doing a good deal of trade.

Chatham Hill, in Rich Valley, near the great plaster deposits of Buchanan's Cove, has two or three stores, a church, and smith shops. It is a rambling hamlet, healthfully situated on the road leading up from the south bank of North Fork of Holston.

Broad Ford, on North Fork, is a place also where a great deal of business is done. And among other known places in the county are, Thomas's, Holston's Mills, Harmon's, Tilson's, Sugar Grove, Chilhowie, Blue Spring, and Sinclair's Bottoms.

WASHINGTON COUNTY.

It is not astonishing that one of the fairest portions of the State should have been chosen to be named after Washington. The county which bears his name is one of the largest, most populous, and among the most important in the State, in every respect. Its large area; wide expanse of grass, grain, and tobacco lands; fine ores, marbles, etc.; fine forests, mineral springs, and noble scenery, make it necessary to use language in its description which, to the impartial reader, sounds like mere fulsome flattery.

Several of the mountain chains, which are so high and rugged farther east, in passing through Washington are so modified as to be no longer the barriers they are in Wythe and Smyth. Thus the Big and Little Walker's Mountains, particularly, have so far disappeared as to present almost one vast plain of undulating fields and woods from Clinch Mountain on the north to Iron Mountain on the south; leaving so much broader an expanse of arable lands to meet the eye, and contribute to the wealth and prosperity of the community. So that in Washington the Great Valley of Virginia is twenty miles wide, unbroken except by inferior ridges, which serve more as divides between the waters of different forks of Holston River than anything else.

Washington, also, holding some of the most considerable towns and villages in that section of the State, claims additional importance on that account.

HOW BOUNDED.

The Clinch Mountain separates Washington from Russell on the north, a length of thirty-three miles. On the south, the Virginia and Tennessee State line is the boundary; toward the southeast is the great White Top Mountain between Washington and Grayson Counties; but that southeastern corner, which is alike the northeastern corner of the State of Tennessee, is not in the White Top Mountain, but on Pond Mountain, seven miles southwest of White Top. East, Washington is bounded by Smyth County, and west by the county of Scott.

HOW WATERED.

The different branches of the Holston River and some of their tributaries afford never-failing streams to every part of the county. A part of the lower section of North Fork of Holston River, as well as the South Fork, discharge enough water to render their improvement for navigation purposes possible. The Middle and South Forks of Holston unite in this county and continue as the South Fork. Laurel Creek, a tributary which derives its waters from streams flowing out of White Top Mountain and from Tennessee, is a stream of great importance, as affording abundance of water at all seasons. It empties into the South Fork. Wolf Creek and others which flow near Abingdon are valuable; likewise is the creek which flows out by Bristol, besides others of value in different parts of the county.

GEOLOGICAL.

It is merely a repetition to put in a geological section for Washington, although for convenience it is proper to do so.

The section for Smyth County shows the different rocks making their appearance in Washington, though there is a difference in their arrangement. Unlike Smyth County, Washington is free from those intermediate ranges known as Glade and White Rock Mountains, and their extension is occupied either by Lower Silurian limestones or St. Peter black shales and brown sandstones, as is the case two or three miles south of Abingdon, in the Knobs.

Toward the extreme southeast are the Huronian rocks, in White Top and Pond Mountains; then going north over the Upper Huronian, the Lower Cambrian are encountered in the north slopes of Iron Mountain; then in the main valleys, as on South Fork of Holston River, Lower Silurian limestones. These limestones then prevail entirely across the valley, except the district of the Knobs just spoken of, and also a line of rather indistinct Upper Silurian rocks near the south side of North Fork, and such of the Devonian series as show the succession of knobs westwardly, toward the Scott County line. On the north side of North Fork are the Devonian rocks in Brushy Mountain, succeeded, northwardly, by Clinch Mountain and the Upper Silurian series. That this enumeration includes valuable mineral-bearing series cannot be doubted.

IRON.

The valuable *Brown Ores* of Washington are mainly confined to the Oriskany measures on the south side of Clinch Mountain.

That portion of Iron Mountain in the county must yield some of those brown ores for which it is noted at other points, but no great developments have been made as yet. Toward the point where the South Fork of Holston flows into Tennessee there are excellent brown ores in the limestones, found close to a band of magnetic ores. From these deposits has been taken a great deal of ore for use in a furnace close by, now out of blast, known as the Eagle Furnace of Sullivan County, Tennessee.

The ores of the Oriskany rocks in Clinch Mountain are not the only brown ores in that section; but they are the most important, not only on account of their greater purity, but more reliable quantity: they, however, follow the rule governing these measures farther east, and are found at intervals along the course of the mountain, on the south side, in a direction about north 70° east, and south 70° west.

Red Iron Ores.

Ores of this class, except a considerable band of pure specular ores on North Fork, are not reported in any appreciable quantity outside of the fossil red ore in the Clinch Mountain. Some handsome fragments have been taken from an 18-inch vein, just east of Little Moccasin Gap, in the Washington-Russell line, on the Fossil Belt.

Magnetic Iron Ores or Semi-Magnetic Red Ores.

The magnetic ores of Washington seem to be of that class, which, at the same time they are abundant, are of the kind most easily reducible. The most considerable deposits occupy a line about 1½ miles south of the Atlantic, Mississippi and Ohio Railroad at the Gollaher Bank, and also running with the general direction of the South Fork of Holston River; and one deposit may be said to show to advantage in the vicinity of the mouth of Fifteen Mile Creek. Here the measure is about three feet—sometimes greater—generally between walls of limestone, dipping at a high angle southwardly.

The Gallaher ore is near a stratum of St. Peter's sandstone, and shows nearly 1,000 tons on the surface.

These measures are more or less continuous for eight or nine miles, with a high probability that farther developments will prove them much longer. The old forge on the south side of the river, near the mouth of Wolf Creek, has used these ores, producing an excellent bar iron.

The next observed line of magnetic ores is on the lands of Preston, three miles east of Bristol, about 300 yards south of the Atlantic, Mississippi and Ohio Railroad. The developments now in progress will reveal their true character, quantity, etc. That the White Top Mountain and vicinity will reveal magnetic ores, specular and brown ores in quantity in this county there can scarcely be room for doubt; though it is now to be regretted that so few developments have been made both in that vicinity and in Iron Mountain.

COPPER

Will be found to exist in several lines of ores in the county, but not in sufficient quantity to justify working on a large scale. At least, up to the present, no developments would lead to other conclusions.

LEAD AND ZINC.

The extraordinary measures which hold such vast amounts of lead and zinc farther east in Wythe, etc., would naturally pursue a line running with the general course of the South Fork of Holston River; but it seems, from the examinations made, that this fine series has been thrown up in the general upheaval to such an altitude as to have been denuded and carried away. No large quantities of either of the minerals may be confidently looked for, to judge from present indications.

PLASTER AND SALT.

It would be difficult to estimate the approximate quantity of the great Saltville deposit assignable to Washington

County. The great fissure in which these invaluable deposits lie extends for some miles into the county, in a direction parallel with and south of the North Fork of Holston River.

It is not at all conclusive that the lesser apparent surface indications determine that a smaller quantity exists in Washington than in Smyth; that is, as to the Saltville deposits. No one can tell what great cavities and inequalities between the sides of the fissure may exist below ground. That there are such the Saltville basin itself is positive proof; and, occurring as it did so near the surface, it needed but little exploration to bring out its character.

Washington County, it may confidently be asserted, holds immense masses and deposits of both these valuable minerals close to the Washington-Smyth line. At the Buena Vista Plaster Works, the quality of the plaster is unexceptionable, and the only wonder is that the sale of this valuable fertilizer isn't treble what it now is.

As remarked in treating of Smyth County, should the owners adopt the plan of showing all its qualities as a permanent fertilizer, as well as a mere stimulant for growing crops, it can scarcely be doubted that the demand for it will greatly increase. This result could not better be secured than by studying thoroughly all the relations borne by gypsum to agricultural chemistry, and then having them thoroughly ventilated before the people of every county in each neighboring State, as well as in Virginia, by a good lecturer at every public gathering.

MARBLE.

A variegated marble in thick bands exists in a line of rocks south of the North Fork of Holston River, and parallel in trend with the general course of that stream. This series may be said to lie nearly at the junction of the Trenton and Hudson sub-epochs. The industrial value of these marbles

has not yet been fully tested; judging from their good
appearance now and then they will be in demand for ornamental purposes.

BARYTES

Exists in some quantity about the middle-southern part of
the county, but the quantity has not yet been fully ascertained by actual development.

TIMBER.

The southern part of the county, in the vicinity of White
Top Mountain and the slopes of Iron Mountain, presents
areas still very heavily timbered with a growth of fine timber.

Poor Valley, except in the vicinity of Saltville, is still heavily timbered. Should the question of quantity arise in connection with the making of iron in the vicinity of the greater
deposits of iron ore, it may be answered that a sufficient
quantity of cheap charcoal is accessible to supply a large
demand for a number of years to come.

The kinds of timber are all those common to the latitude.
In the southern part, besides balsam, lashhorn, etc., there
are quantities of wild cherry, poplar, etc. About through
the great valley district there are fine walnut trees, and in
places quite abundant.

White oak is the prevalent tree. In the Clinch Mountain
are large boundaries of chestnut, chestnut oak, with hickory,
etc.

WATER POWER.

Washington could afford water power for any desirable
purpose if called upon. The different branches of Holston
River and their tributaries offer facilities possessed by only
a few counties in the State. Taking into consideration the
large area of the county, and the unfailing character of the

rather large streams contributing power, the aggregate number of mill sites which may be used indifferently for large grist-mills, cotton or woolen mills, or saw-mills, is great enough to defy computation.

South and North Forks, where they flow into Tennessee, discharge about 400 cubic feet per second each.

AGRICULTURE.

Agriculture being the chief pursuit of the people, and the land being in a fine grass-producing section, the county now derives nearly all its revenues and support from that source. Like its sister counties, Washington does not pretend to have made much headway in farming as a science; land is still too abundant and cheap, and population too scarce; but nature has made these limestone lands so rich, and has given such propitious seasons, together with so admirable an elevation above the sea, that the different grasses, growing in rich profusion season after season, would almost alone render the county remarkably capable of producing the best and most constant revenue.

Space would not permit a description of particular localities most famous for the fertility and productiveness of their soil. It would be an invidious task at best. To get on some elevated point and look over the broad expanse of the county, the eye meets with a most pleasing picture in the alternation of hill and dale, of woodland and pasture, occupying such a widely extended area. It is not a mere series of plantations, side by side along the banks of a large stream, with all the rest in swamp or inaccessible mountain; the whole broad valley is quilted over with farms of princely size, and many of them of surpassing beauty.

In wheat, corn, grasses, rye, oats, barley, buckwheat, flax, etc., the county for the greater part has few equals as a producer.

TOBACCO CULTURE.

Of late years tobacco culture has reached a high limit in the county, and that staple is now being produced to the extent of 1,200,000 pounds of leaf annually.

MINERAL SPRINGS.

The Seven Springs, between Glade Spring and Saltville, on the Saltville branch of the Atlantic, Mississippi and Ohio Railroad, is the place at which is now being made the widely and justly famed Seven Springs Iron and Alum Mass. At these springs are nicely arranged furnaces and boilers for reducing the water drawn from the Seven Springs, the waters from which, when analyzed together, gave in advance the medicinal constituents which have proven of such high efficacy in nearly all forms of disease. DR. J. W. MALLET, of the University of Virginia, is the chemist whose searching analysis first showed the wonderful therapeutic value of this water.

It is now confidently believed that, with an energetic system of advertising throughout the country, the demand for this mass must be far in excess of any possible supply.

A most careful inquiry into results reveals the fact, that out of the thousands of medicines of every conceivable kind put before the people as infallible, the mass of these springs has been found to yield an almost infallible remedy for the diseases indicated by the distinguished chemist named above.

Mungel's Springs, situated nine miles northwest of Abingdon, has a fine spring almost in the edge of the North Fork of Holston River. This spring yields a distinctly white sulphur sediment, with an arsenical tinge. It has a high local reputation for curative virtues, and, with proper accommodations for visitors, should command a good patronage. Its situation is romantic and picturesque to a very high degree, with a lovely river flowing through rich scenery, set in a background of high hills overtopped by higher mountains.

WASHINGTON SPRINGS.

These springs are one and a half miles north from Glade Spring Depot, Atlantic, Mississippi and Ohio Railroad. Besides being situated in a lovely spot amid the mountains, with extensive views of noble plains and the vast mountains about the White Top, its springs are justly regarded as distinctly medicinal and of high curative power. At these springs are found waters of four distinct varieties, the most effective being an alum-chalybeate spring near and east of the hotel, and a white sulphur spring, yielding a low percentage of arsenic, situated in a lovely spot west of the hotel. If mere curative power is a matter of importance, these springs may be ranked among those most likely to sustain a high reputation. To this should be added the healthful and beautiful location close to a leading line of important railway.

Mendota has in its vicinity, in the Hamilton Slates, several good sulphur and chalybeate springs. In the county, at numerous places, are springs of lesser note, needing only development to prove their efficacy and value.

SCENERY.

When it is remembered that from nearly all parts of the county can be seen the White Top and Balsam Mountains, towering 5,500 feet above the sea, with other grand mountains in the distance, it is not too much to say that Washington County presents every description of fine scenery. A view of the vast plain of the great valley itself, in its garniture of mountains, is most beautiful and pleasing. Such of

ARSENIC SPRING, WASHINGTON SPRINGS, WASHINGTON CO., VA.

(P. 126.)

the creeks as lead down from the higher mountains, with
their bright limpid waters dashing over numerous ledges and
boulders, in cascades and falls of a thousand different forms,
fringed by dark foliage composed of tree and shrub, present
innumerable pictures at once romantic and surpassingly
beautiful. No power can describe the inimitable view from
the White Top Mountain. From this elevated point the distant mountains of Kentucky are visible far to the northwest.
The serried lines of the parallel chains of Virginia mountains
stretch away to meet the sky, until the view is lost in the
azure haze of the great distance. To the south, the lone and
lofty mountains of the Unaka Range relieve all sameness, not
more by their isolated grandeur, than their beautiful and gigantic proportions.

MANUFACTURES.

Washington County has shown a fine spirit in the establishment of fine woolen mills, so well calculated to consume
at home its own surplus wool, and much of that of the surrounding counties. Besides these woolen factories and the
large tobacco factories at Abingdon and Bristol-Goodson,
there are no manufactories of consequence. One mile west of
Abingdon, on Wolf Creek, are situated the woolen mills of
J. H. Pepper & Sons, running one set of cards, 340 spindles,
and 4 looms, making jeans, flannels, linseys, cassimeres, and
blankets. A regular custom mill, run by a Leffel's turbine
wheel.

The Bristol Woolen Mills, half a mile east of Bristol, on
the Town, or Beaver Creek, runs 504 spindles and 8 looms by
water power—most improved 48-inch machinery—making
linseys, cassimeres, jeans, satinets, and blankets, consuming
50,000 pounds of wool annually.

It may not be out of our province to mention the mills on
the Tennessee side of Bristol. The City Woolen Mills are

situated on the same creek, half a mile west of Bristol, using 40-inch machinery, run 240 spindles and 5 looms, consuming 100 pounds of wool a day. Goods of excellent quality, cassimeres, satinets, flannels, blankets, shawls, etc.; yarns card and spun for farmers. Business chiefly, like that of Bristol Woolen Mills, with the farmers, in exchange, at the rate usually of one yard of satinet to two pounds of top-washed wool; cassimeres, one yard for two and a half to three pounds of top wool.

Below these mills, about one mile, are the Bristol Cotton Mills, on the same stream, 992 spindles, 18 looms; product, 80 bunches of yarn daily, and 650 yards of sheeting. Factory employs twenty-five hands. One half of the machinery was made by Danforth, of Paterson, N. J. The new, or last half, came from the machine works of Lowell, Mass.; appears very fine indeed.

At Abingdon is situated one of the finest tobacco factories in the State. The quality of the article produced bears favorable comparison with the product of the older factories. It was established in 1876, and now handles over 1,500,000 pounds annually. At Bristol-Goodson there is also a tobacco factory, with the prospect of another soon. There are also at Bristol two tobacco warehouses handling about 1,000,000 pounds of leaf—possibly 100,000 pounds of chewing and smoking tobacco. At Abingdon the tobacco warehouses are the Greenway, Snow, and Holston factories, to which has just been added another.

FURNACES AND FORGES.

Washington County has now no iron furnaces in operation. There is a good forge of 700 pounds capacity, situated near the mouth of Wolf Creek, on the south bank of the South Fork of Holston River, which runs chiefly on magnetic ore of that vicinity, not now in blast.

Abingdon, the county site, with a population of 1,700, is one of the oldest towns west of the Blue Ridge. Its situation is pleasant if not beautiful, having many attractive features about it.

Its two fine female colleges, handsomely situated, add much to the attractiveness of the place. Abingdon, besides these schools, has a spacious court-house in which is held, not only the county and circuit courts of the commonwealth, but the circuit court of the United States for the large district of which Abingdon is nearly the geographical center. Here are also churches of nearly all denominations, three well-kept hotels, a good livery stable, numerous stores, dealing in every description of merchandise, medicine, stationery, etc. There are tanneries, establishments for the manufacture and repair of wagons, harness, smith shops, etc. This town being situated nearly in the center of the county, on the Atlantic, Mississippi and Ohio Railroad, commands a considerable trade, not only from the county of Washington, but from surrounding counties in Virginia, Tennessee, and North Carolina. Among Abingdon's chief institutions are her two enterprising weekly public prints, *The Standard* and *The Virginian*.

BRISTOL.

Bristol is a town of 4,000 inhabitants lying in the States of Virginia and Tennessee, at the western terminus of the Atlantic, Mississippi and Ohio Railroad; its Virginia portion being usually known by the name of Goodson.

It is a town of quite recent origin, dating back to about 1858. Bristol is the center of quite a manufacturing district; besides the woolen, cotton, and tobacco factories enumerated above, it has quite an extensive machinery for facilitating carpenter's work.

9

Its enterprising citizens have made Bristol the center of a large and growing tobacco trade; in fact there is no branch of industry left neglected. The various newspapers published there, of which there are two or more, have been very efficient in building up the place, though it has not had such a magical growth as most of Western towns. Its hotels are sufficient in number and attractiveness to be efficient aids in the development of the place. The stores of all kinds of merchandise, attractive watch-making establishments, presided over by the best talent in the country, together with churches of various denominations, make Bristol a place of note in the surrounding country.

GLADE SPRING.

This name applies to two places near each other, Glade Spring Depot, at the south terminus of the Saltville Branch Railroad, taking its name from Old Glade Spring, which is situated two miles to the south on the old stage road.

Glade Spring Depot is an inviting looking place, with most of its houses built in good style and freshly painted. Its good hotel helps to render it a desirable place to spend the hot summer months. The cool vine-covered veranda, good table, and cleanly rooms will be remembered by many who have partaken of its comforts. This place has some trade, chiefly with the southern part of the county, Tennessee, and North Carolina. It is supplied, like nearly all the towns, villages, and hamlets of Southwestern Virginia, with Masonic or Odd Fellows lodges, or both.

BUENA VISTA.

Buena Vista is in that part of the Saltville Basin extending into Washington County. It is from its mills that the excellent article of plaster known as Buena Vista plaster comes.

EMORY.

Emory is the noted station, on the Atlantic, Mississippi and Ohio Railroad, for Emory and Henry College, through the grounds of which college the railroad passes.

On this railroad west of Abingdon are *Wallace's* and *Montgomery's*, two points that bid fair, at some day, to become trading places of some note.

Mendota is a village on the north side of North Holston River, important on account of its good school, besides having some trade with the neighboring country.

Greendale, *Friendship*, and *Mock's Mill* are places of some note in the county.

LINES OF TRANSPORTATION.

The ATLANTIC, MISSISSIPPI AND OHIO RAILROAD passes through the heart of the county, in one of the great through lines from New York to New Orleans. From Glade Spring starts the Saltville branch of the Atlantic, Mississippi and Ohio Railroad, nine miles to Saltville. From Bristol to Cumberland Gap is now being constructed a narrow-gauge railroad —known as the Bristol Coal and Iron Railroad—by the Tinsalia Coal and Iron Company, the object of which is to bring Bristol and connections in communication with the vast beds of coal and iron along the route of the proposed road.

Washington County will also derive a proportionate advantage from the construction of the Saltville and Coal Mine Railroad, the route for which is now being surveyed.

Both forks of Holston River might be determined, upon a close examination, to be susceptible of being made navigable.

FISH CULTURE.

Some gentlemen of the county are taking quite an interest in the propagation of fine varieties of fish. The German carp seems to be a favorite. No doubt the State Commission will take the waters of the Holston under its special care soon. Bee culture is an industry of great local value, as evidenced by the interest taken in different improved hives. Grape culture in varieties for home use is carried to considerable perfection.

ANNUAL SURPLUS OF CATTLE, SHEEP, WHEAT, CORN, TOBACCO, ETC.

Fat cattle, 2,500 head.
Stock cattle, 5,400 head.
Sheep, 9,300 head.
Wheat, 60,000 bushels.
Corn, 1,000 bushels.
Oats, 5,000 bushels.
Leaf tobacco, 465,500 pounds shipped.
Manufactured tobacco, 130,000 pounds from Abingdon, and about the same quantity from Bristol-Goodson Factory.
Staves, 3,362,000 pounds.

It is almost impossible to get at the exact number of cattle and sheep in any county; but the above estimate, made from different sources, may be said to give reliable figures.

EDUCATION.

Much to the credit of the citizens of the county the subject of education has always been one of great importance, and has fully succeeded in engaging their attention to good effect.

Outside of the public schools there are no less than five permanent colleges and schools, four of which are institutions chartered to grant diplomas for a full course of scholastic learning.

Emory and Henry College is the principal male college,

and Martha Washington College and Stonewall Jackson Female Institute, both of Abingdon, are female colleges of high merit.

EMORY AND HENRY COLLEGE,

Situated twelve miles east of Abingdon, on the Atlantic, Mississippi and Ohio Railroad, was established under authority of the State, in the year 1838.   It has had quite a successful history in the past, some of the most distinguished men of the country having been students there.   With an excellent faculty now, and fine facilities for education, it is entitled to a large and growing patronage.   The grounds and buildings, laid out with great skill for the objects aimed at, have been brought in the course of over forty years to a high state of beauty as well as adaptation to the purposes in view.   It has a very attractive feature, also, in a fine farm of over 300 acres.   From the beauty, convenience, and perfection of this college, in all its appointments, we are led almost irresistibly to advocate still further the views set forth in the treatment of the County of Montgomery—namely, that Emory should be secured by the State, and turned into the Agricultural and Mechanical College, combining the excellent faculties of these two institutions, and using the buildings at Blacksburg for the purposes indicated in the remarks on that part of Montgomery County.

Without going more into detail in the description of Emory and Henry College, the subject may be dismissed with the hope that the excellent and convenient location of the college, its fine faculty, and the really beautiful arrangements of its buildings in a place of great natural beauty and healthfulness may bring it a prosperous future.

MARTHA WASHINGTON COLLEGE

Was established at Abingdon, previous to 1861, to be a college, of high grade for young ladies.   It is generally be-

lieved to fully meet the expectations entertained of it by its
friends.

Its faculty is highly recommended by the most experienced
talent in that line in the State, and it is fair to presume that
this institution is justly entitled to high commendation.

It occupies a beautiful and tasteful building surrounded by
admirable grounds in the old town of Abingdon.

STONEWALL JACKSON FEMALE INSTITUTE

Is also situated in Abingdon, with grounds and buildings that
should alone speak volumes in its favor. It was established
subsequent to 1865. Its history has been a record of a degree of success which its friends could hardly have hoped
for it. Its name, its painstaking faculty, and fine situation,
should enable it to command a growing patronage.

Sullin's Institute of Bristol, though just over the line in
the State of Tennessee, is regarded almost as a Virginian
institution. It has been very largely patronized by young
ladies from Virginia, and bids fair to have a very successful
future. The same may be said of King's College, likewise
located at Bristol. *Mendota High School for Boys and Girls*
is regarded as one of the best schools of its kind in that
section. It has had, hitherto, great success, and has drawn
pupils from quite a distance.

There are other excellent schools in different parts of the
county, besides which are the usual number of public schools
throughout the county, now reported by the Superintendent
of Public Education as increasing in efficiency.

GILES COUNTY.

It is rare that nature repeats such a combination of fine
ores in veins and deposits, mineral waters, superior grass
and grain lands, together with noble forest, river, and lake

scenery, as is presented in the area covered by Giles County.

The great mountain chains, which may be said to occupy the southwestern prolongation of the Alleghany Range, are here broken in two, and apparently swept back, like the stately and beautiful structures which form the sides of a pair of gates incomparably great in size and architecture. In the great basin formed by this vast opening is the heart of the county: green fields and forest-covered hills, threaded through the center by the beautiful and rapid New River ; a stream buttressed half its length by lofty cliffs of limestone, carved out, in the course of time, into such shapes as to lift the scenery along it out of the mere commonplace, and elevating it into the beautiful, not to say the sublime.

To its other more notable features, Giles County adds the highly important one of being the great gateway of all the projected lines of railway, both leading from the Virginia seaboard toward the great west, and from north to south. In this particular this county is peculiar, seeming to occupy a position which brings it within an air line for four different east and west railroad lines, and for two from north to south.

To do justice to this noble county, in the space here allotted to it, is an impossibility. Indeed it is a question, whether or not any description could be written adequate to the just claims of this county to pre-eminence, considered in all its features actual and possible.

HOW BOUNDED.

Giles County is now one of the border counties of Virginia; its northern neighbors being the counties of Monroe and Mercer, in West Virginia, separated from Giles by the great iron-bearing mountain range known as Peter's and East River Mountains.

West, it is bounded by the county of Bland, Va., east, by
Craig County, Va., and south by the counties of Montgomery
and Pulaski, two of the important counties treated in this
volume.

HOW WATERED.

The whole area of the county is well watered by New
River, flowing through the middle of the county, from south
to north, and several of its larger tributaries, such as Big and
Little Stony, Sinking and Doe Creeks, on the east side, and
Wolf and Walker's Creeks, with minor tributaries, on the
west side. Little Stony Creek has for its source the celebrated mountain lake on the top of Salt Pond Mountain,
4,000 feet above sea level.

NOTABLE PHYSICAL FEATURES.

As before remarked, the northern boundary line is marked
by Peter's and East River Mountains, really continuations of
each other, Peter's Mountain being to the northeast of New
River, and East River Mountain to the southwest, in the
same line. Next toward the central part of the county is the
lofty and beautiful Angel's Rest, about 4,000 feet above sea
level; opposite to which, on the northeast side of the river,
is the Butte Mountain, of the same general elevation. Flanking this latter on the south is the Salt Pond Mountain, with
its bald knob towering nearly 5,000 feet above the sea; the
northwestern face of this high knob being washed by the
crystal waters of Mountain Lake, no less remarkable for its
beauty and elevated position in the top of a high mountain
than for the fact of its recent origin.

The Salt Pond Mountain, on the northeast, seems to
answer in position to the Sugar Run Mountain, on the southwest side; leaving between them, as well as between Angel's Rest and Butte Mountain, six or eight, or more miles

Section through Giles County Va.

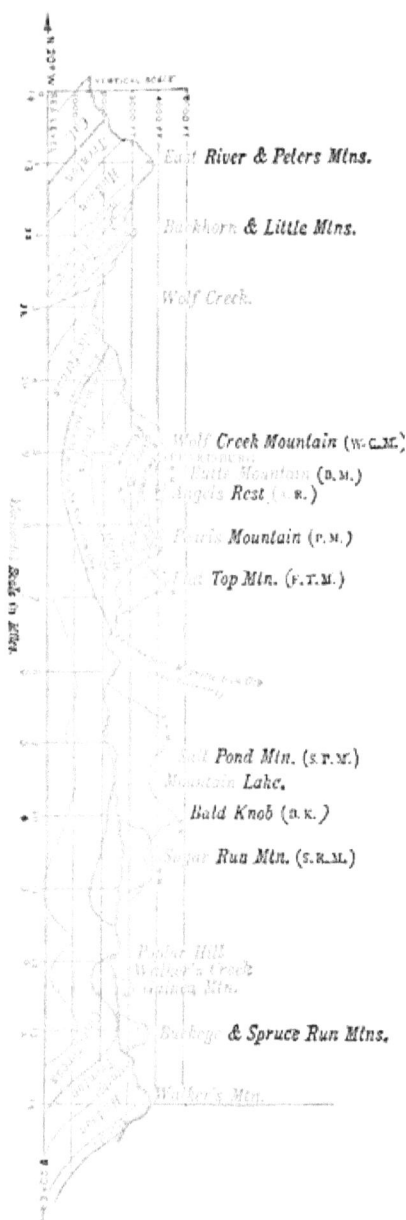

NOTE.— The Geology of Butte and Salt Pond Mountains is nearly identical with that of Angel's Rest Mountain. This section being taken west of the Buckeye and Spruce Mountains, it would be difficult to show the order of stratification and dip of the rocks in those mountains without destroying the proper order of the section.

East River & Peters Mtns.

Buckhorn & Little Mtns.

Wolf Creek.

Wolf Creek Mountain (w. c. m.)
Butte Mountain (B. M.)
Angels Rest (a. r.)

Pearls Mountain (p. m.)

Flat Top Mtn. (f. t. m.)

Salt Pond Mtn. (s. p. m.)
Mountain Lake.
Bald Knob (B. K.)

Sugar Run Mtn. (s. r. m.)

Poplar Hill
Walker's Creek
Spring Mtn.

Buckeye & Spruce Run Mtns.

Walker's Mtn.

of limestone grass lands, divided in the middle by New
River.

Toward the southern side of the county are the important iron-bearing parallel series, composed of Spruce, John's
Creek, and Gap Mountains on the northeast side of New
River, and Buckeye, Guinea Mountain, and Walker's Mountain on the southwest side of the river, Gap Mountain
and Walker's Mountain answering to each other in line of
continuation.

Angel's Rest Mountain is but the northeastern terminus
and culmination of the great iron-bearing ridges of Wolf
Creek Mountain, Pearis Mountain, and Flat Top; to the importance of which the reader's attention will presently be
called.

GEOLOGY.

The geology of Giles County rocks is comprised between
the Upper Calciferous limestones, and the Hamilton Black
Slates, inclusive. In the latter are sometimes found two or
three inches of impure bituminous coal, creating the impression in the minds of the uninitiated that there are valuable
beds of coal close under the surface; but this hope will not
be realized.

To properly illustrate the geology, there should be drawn
several sections across the country, from northwest to southeast; but it is hoped that the section here given will be
ample to show the positions of the various strata relative to
each other, as well as their general position, dip, etc.

Beginning at the southern or southeastern end of the section, the rocks of the Oneida series are first encountered,
dipping southeastwardly, at angles varying between 30° and
60°, and occupying a position nearly in the heart of Walker's
and Gap Mountains; being on the extreme southern boundary line they dip at once out of the county, just here.

Next to the north of, but under the Oneida, are the variously colored rocks, highly impregnated with lime, belonging to the Hudson series, nearly 1,000 feet thick; next to the north, the outcrop of some of the Trenton limestones, 850 feet thick about. Here a fault or great plication is encountered at the northern base of Walker's Mountain, and, at the south base of Buckeye Mountain, the Oriskany rocks are encountered, about sixty feet thick, showing six to eight feet of fine brown ore, both in Buckeye and Spruce Run Mountains; then north of this the red sandstones and ores of the Clinton; then the Oneida, bounded on the north by the Hudson series again, with its various colored limestones, based on thirty feet of fine variegated marble, somewhat similar to Tennessee marble; then the Trenton, 500 feet thick, gradually losing its steep angle of dip as you near the great basin in the heart of Giles County, and becoming, for considerable distances, almost horizontal. Near the bottom of the Trenton rocks, if not in the division between them and the Calciferous series, is the position of the famous Giles County semi-magnetic red iron ore.

This vein of ore is one of the most remarkable ever encountered. It has been a source of much speculation in the minds of scientific men, and is yet, in the estimation of many, an undetermined problem. It deserves a chapter to itself, and will receive more full notice further on.

Pursuing the section northwardly, you begin to ascend out of the Trenton into the Hudson series, showing in the base of Angel's Rest Mountain, as well as in a corresponding position in Butte and Salt Pond Mountains. This is then overlaid in regular order by the Oneida, Medina, with the Clinton capping the summits of these last-named mountains; excepting some very large areas, which have also the intervening rocks up to the Oriskany, inclusive. In these last-named mountains, as will be observed in the section in Angel's

Rest, the rocks dip gently toward the center of the mountain.

Leaving these great mountains, going north there is a dislocation about the line of Wolf and Big Stony Creeks, and you then encounter the steeper dips on the south faces of Buckhorn, East River, and Peter's Mountains, and their flanking south-lying ridges. In these all the rocks from the Hamilton Black Slates to the Oneida, inclusive, are found, dipping southwardly at angles varying between 35 and 60.

The writer here, in closing the geological description of Giles, pays a justly deserved tribute to the fine discrimination shown by Prof. Wm. B. Rogers while in charge of the geological survey of Virginia. All subsequent work has proven him correct to a degree surprising for the small amount of development at the time of his explorations.

IRON ORES.

The iron ores of Giles County may be divided into three general classes, as follows: The Semi-magnetic Red, the Fossil Red, and the Brown Ores or Limonites. The semi-magnetic red ores showing at Johnson's, Chapman's, and Pack's, on and near New River, occupy a position, as mentioned above, nearly at the junction of the Calciferous and Trenton limestones, possibly just between the Black River and Trenton sub-epochs. At Johnson's, on the river, the deposit has been very well explored, and found to give a thickness there of over fifty feet in different seams. A few hundred yards up the river from this point the New River Railroad Company, with the diamond drill, has ascertained its continuity beyond conjecture, finding nearly 100 feet of ore in a little over 350 feet, some of the seams attaining a thickness of 18 feet.

But the strata in which this ore is contained show very distinctly near Newport, at Payne's, twelve miles away from

Johnson's and Chapman's, at Moser's, and at the mouth of Big
Stony Creek ; then on the west side of the river, besides
Pack's, at Dill's, Jordan's near Dill's, Wood's, Eaton's, near
Wabash Camp Ground. All these latter places occupy a
position in the rim of a great basin, having its greatest depression near Johnson's or Chapman's Ferry. Again this
same stratification shows along the south side of Wolf Creek,
on the lower slopes of Wolf Creek Mountain, and around the
northern base of Angel's Rest Mountain, under which mountain the whole stratification lies in its original strength. It
is a peculiar vein. Among its more noticeable peculiar-
ities is its being cut out, now and then, by slaty material
sometimes highly charged with iron; that is, it is really
replaced in the walls by slates, and sometimes by chert.
The ore is usually of slaty structure, of tabular form sometimes, generally though lozenge or rhombus shaped. It
may be submitted that, in the hearts of the hills and below
water, the ore has not been replaced by any other material,
and may there be regarded as much more regular and reliable, as has been proven by the drill. It is true that this
vein sometimes shows red and brown ores accompanying the
more magnetic; but this is only due, very probably, to local
causes. The general stratification in which it occurs is
marked above by felspathic chert 10 feet thick, which occupies a position about 75 feet above. Below it occasional
lumps of *baryta* are found.

This remarkable stratification dips under Angel's Rest,
Pearis's and Wolf Creek Mountains ; on the other side of the
river it disappears under Butte and Salt Pond Mountains,
while toward the south side of the county it goes under
Buckeye and Spruce Run Mountains.

It may not be uninteresting to state that at Johnson's,
below Chapman's Ferry, the exposures made show a section
of the vein's area 220 feet by 170 feet by an average thick-

ness of eleven feet. Taking the ore to weigh 280 pounds to
the cubic foot, we have here, practically in sight, of this
famous ore about 51,500 tons. The surface outcroppings, as
described above, lead to the conclusion that there are many
repetitions of the section showing at Johnson's, though at
that point the trough of the basin is below water level.

An analysis by Dr. F. A. GENTH is as follows:

Moisture.......................... 0.12 per cent.
Silicic acid .... 1.83   "
Titanic " .... none.
Phosphoric acid................... none.
Sulphuric   " .... 0.05 per cent.
Magnetic oxide of iron ....71.36   "
Ferric oxide ....26.52   "
Alumina ....: trace.
Magnesia.......................... 0.07 per cent.
Lime.............................. 0.05   "

───────
100.00
Metallic iron..................... 70.238 per cent.
(Signed)   F. A. GENTH.

Fossil Red Iron Ore.

The Clinton or Dyestone series in which this ore is contained, together with the Medina, measures about 200 feet;
but the fine ore is never, in this county, over three feet thick.

Pearis's Mountain, Wolf Creek Mountain, Flat Top, Sugar
Run, East River, Buckhorn, and Buckeye Mountains, west of
New River, and Peter's, Little, Butte, Salt Pond, Spruce
Run, and John's Creek Mountains, east of New River, are its
localities. It may be said to possess the same general character all over the county where it is found. It has a fine
development both in Flat Top, Buckhorn, and East River

Mountains, in its walls of red sandstone, standing at right angles, and generally from 800 to 1,000 feet above water level at its outcrop. Its quantity, estimating the average thickness of the best ore at 18 inches, would be greater than that of all other ores combined in the county, from its great regularity and continuity.

Some fragments of this ore found in the great mountain area west of the Angel's Rest are very beautiful as cabinet specimens; frequently a perfect shell will be found petrified in specular ore of the brightest silvery luster. Generally the fossils are small, flattened, and lenticular shaped, mingled with small rounded pebbles.

Two analyses, as rendered from fossil ores taken in East River and Flat Top Mountains, are as follows:

East River Mountain Ore.

Metallic iron 50.36

(Signed)   Prof. FESQUIT.

Flat Top Mountain Ore. (Fossil Red.)

Sesquioxide of iron 58.12
Oxide of manganese 0.06
Alumina 4.66
Lime 0.20
Magnesia 0.41
Potassa and soda 0.40
Silica 32.74
Sulphuric acid 0.00
Phosphoric acid 0.75
Water, hygroscopic 0.69
Water, combined 0.96
Organic matter 0.84

(Signed)   H. DICKINSON,
                        Norwood, Mass.

Brown Iron Ores.

The quantity of Brown Iron Ore above water level in Giles County would be very difficult to approximate. That it is exceedingly abundant no one would doubt who could see the great Chestnut Flat Deposits, the ore banks from which the John's Mountain Furnace, near Newport, derives its ores, as well as many other notable places in the county. The most conspicuous and valuable beds are found in the Oriskany rocks in Wolf Creek Mountain, East River, Buckhorn, Peter's, Flat Top, Sugar Run, Buckeye, Butte, Salt Pond, John's Creek, and Spruce Run Mountains, and their secondary or derivative deposits on such creeks as Big Stony, etc.

This ore constitutes about ten per cent. of the rocks composing the Oriskany measures, which are generally from thirty to sixty feet thick throughout the localities above named. The Chestnut Flat ore is really red when crushed. It is altered from a brown ore. This is amply proven by following the same measures a short distance either way, when the ore is found to resume its character as a brown ore.

Chestnut Flat, with its fine showing of ore on the crest of Wolf Creek Mountain, about three and a half miles, air line, southwest from the narrows of New River, is one of the most remarkable places in Virginia. This singularly fine ore is thrown up over a distance of about three hundred yards in an almost solid body, but extends less conspicuously many hundreds of feet further.

It will probably yield 300,000 tons, just at this point, of an ore which analyzes as follows :

Sesquioxide of iron.......89.65 = Metallic iron, 62.755.
Oxide of manganese...... 0.20
Silica.................... 2.58
Alumina................ 1.11
Lime 0.20

Magnesia............... 0.15
Sulphuric acid.......... 0.37
Phosphoric acid.......... 0.30
Water, hygroscopic....... 1.25
Water, combined......... 4.10

(Signed)   H. DICKINSON,
Norwood, Mass.

Following this same lead westwardly about one and a half miles, along the south face of Wolf Creek Mountain, looking down upon No Business Creek, another very extensive deposit is encountered, nearly if not quite as great in quantity as that at Chestnut Flat.

This ore is a brown ore, giving the following analysis by the same chemist :

Sesquioxide of iron...........86.17 = Met. iron, 60.669.
Oxide of manganese .... .07
Alumina.................... .69
Lime...................... .14
Magnesia .... .06
Silica..................... 2.10
Sulphuric acid .... .33
Phosphoric acid............. .46
Water, combined........... 8.01
Water, hygroscopic.......... 1.10

(Signed)   H. DICKINSON.

Again, about three miles almost directly south of this, ores of the Oriskany show in the south face of Flat Top Mountain, overlooking Dismal Creek, being the same from which the Walker's Creek Forge once obtained its supplies of ores.

This ore assays as follows :

Sesquioxide of iron...........80.17 = 56.119, Met iron.
Protoxide of iron............. .57

10

Silica............................. 5.13
Alumina........................... 2.04
Lime.............................. .47
Magnesia.......................... .14
Oxide of manganese................ .90
Sulphuric acid .... .02
Phosphoric acid .... .84
Water, hygroscopic................ 1.60
Water, combined 7.94

                         (Signed)   H. DICKINSON.

Several other places on Dismal Creek give this ore in vast quantities, particularly toward the head of the creek, where the material has not been so abraded and carried away through the action of the elements. No doubt the most convincing argument respecting its quantity would be for the curious or the doubting to visit these beds. Exact measures are superfluous where nature has been so lavish.

Another prominent locality of this Oriskany ore is in the end of the south flank of Peter's Mountain, looking down upon the Narrows of New River. It is supposed by some experts that there are here about 100,000 tons of the ore practically in sight. Its analysis shows it high in metallic iron, and low in phosphorus.

In Spruce Run Mountain, west of Newport, and in Buckeye Mountain, the solid ore is six feet thick, running for great distances, that is, almost continuously. Occasionally the ore is eight feet thick between walls of sandstone.

At Dowdy's, Mill's, Keffer's, etc., you find the eastern prolongation of the Spruce Run veins. These ores may be seen at the furnace, above Newport, on Sinking Creek.

Again, brown ores show us probably the result of decomposed pyrites in the edge of Wolf Creek, near Shumate's house and mill—vein perhaps five feet thick between ledges

of limestone—at Bolton's, up Big Stony Creek, at points on
Guinea Mountain, and numerous other places in Giles County,
to which the above enumerated may be considered as merely
introductory.

MANGANESE.

Manganese ores seem to be confined almost exclusively to
the Oriskany measures. In fact, the iron ore of those rocks
frequently gives way almost entirely to oxide of manganese.

At one point in these rocks on Flat Top Mountain, near the
line between Giles and Bland counties, the ore was found in
great purity, giving the following measures, etc.: Trend north
70˚ east, dip 60˚ north 20 west, containing valuable quantities of manganese disseminated heavily through sandstone,
five hundred yards in length, gradually becoming impregnated
with iron as you approach the eastern end. The apparent
width of the ore strata is here extraordinary, and may be
owing to a duplication of strata from end pressure, or flexure,
or a mere fold. It is 240 feet through. Elevation of outcrop
above water level in Kimberling Creek is 1,200 feet; vein
would no doubt strip well.

Analysis of manganese ore as follows:

Analysis of Manganese Ore.

Red oxide of manganese.......................84.34
Oxygen 3.73
Protoxide of cobalt..........................   .68
Alumina...................................... 1.80
Lime...   .32
Silica.......................................   .21
Baryta 7.21
Water 1.71

                    (Signed)   H. DICKINSON.

148   GILES CO.—LEAD AND ZINC.

(This analysis being made out by Mr. Dickinson in close technical form, gives 59.215 p. c. of metallic manganese, 23.53 p. c. of combined oxygen, and 3.73 of oxygen recognized as otherwise disposed in the ore.)

Some of these ores resemble closely the chalcophanite, so admirably investigated by E. S. DANA and by Dr. BROWN of Liberty Street, New York.

Again, manganese ore shows in beautiful crystals of pyrolusite up Big Stony Creek, as common oxide in Buckeye and Spruce Run Mountains, and at the ore beds above the furnace on Sinking Creek. It is not yet fully determined whether these ores will pay as shipping ores, either for quality or quantity. In the Salt Pond Mountain considerable masses and deposits have been found, that is, in the flanking ridges to the south. It is probable that the bright silvery luster sometimes characteristic of the ore has deceived some people into mining and shipping it, under the impression that they had good silver ore. Such is presumed to be the case with reference to ores shipped from near Salt Pond Mountain.

COPPER.

Copper has been detected in the pyrites of the Hamilton Black Slates, but in small quantities.

LEAD AND ZINC.

Both lead and zinc are occasionally met with in this county. The rocks of the Lower Helderberg group sometimes give them in small quantities. These Helderberg rocks are usually in the form of very pure limestones, rarely ever more than 50 feet thick. Dismal Creek shows these rocks with lead in them. They may also be found at a few places on East River and Buckhorn Mountains, Big Stony Creek, and now and then in the Butte and Salt Pond Mountains.

SILVER.

A trace of silver may now and then be found in the lead ores just described, but as to any paying quantity ever being discovered it is quite doubtful.

LIMESTONE.

Limestones abound in nearly all parts of the county, except in the higher mountains. They belong usually to the Lower Silurian or Cambrian Series. Only those few ledges above mentioned as containing the lead, which belong about the division between the Upper Silurian and Devonian Series, are out of the usual locality occupied by the great mass of limestones in the county.

Along New River, all the way through, except in the mountain cañons, the limestone forms many towering and beautiful cliffs.

An analysis of some of the Helderberg limestone, taken from Dismal Creek by PROF. DICKINSON, is as follows:

Lime 49.42
Magnesia 2.04
Protoxide of iron 1.53
Oxide of manganese 15
Alumina 48
Silica 2.91
Sulphuric acid 02
Phosphoric acid 04
Carbonic acid 42.00
Water 60
Organic matter 78

There are a great many ledges of pure limestone in the county, making an excellent lime, while occasionally hydraulic limestone is met with. The first weathers blue, the latter light drab or earthy white.

MARBLE.

There is a ledge of very compact limestone, about 30 feet thick apparently, at the junction of the Trenton and Hudson Rocks, very full of the remains of comminuted shells, etc., which may very well be considered a fine marble. In many places it resembles the Tennessee marble in appearance and texture. Its best development is probably in Rye Hollow on the south side. This locality is toward the south side of the county, west of the river. It shows also on the east side of the river, in the north of Spruce Run Mountain, low down, and at other places, including the north and south of Angel's Rest Mountain, Butte and Salt Pond Mountains.

TIMBER.

Timber is very plentiful in Giles County. Much of its area would yield largely for years to come in charcoal. There is white oak, chestnut oak, chestnut, black oak, hickory, sugar maple, locust, black pine, white pine, hemlock or spruce pine, poplar, linn, buckeye, black walnut, dogwood and cedar, in the order of their respective quantities.

Much of this timber is very fine for cabinet and ornamental purposes. Large areas of the mountain sides yield immensely in chestnut oak, from which tan bark could be obtained in such quantities as to make a paying industry, once the question of transportation is solved.

In the valleys and among the strictly grass lands, nearly all the trees except the white and black pines, chestnut oak, hemlock, etc., abound. Some of the walnuts and sugar trees attain remarkable size and beauty. Many of these noble trees, which would bring a high price in the seaboard cities, have been burnt in log heaps, or have been used in making worm-fences.

AGRICULTURE.

It is difficult to do this subject justice in a few words, as far as it relates to this county. The sections of the county which are most favorable for farming are those which are also the best grass producing sections. While this is true of nearly the whole of Southwestern Virginia, in the limestone counties, it seems to be particularly so with regard to Giles. The great central basin on either side of New River, ten or twelve miles in average width, Upper Sinking Creek, Walker's Creek, Sugar Run, Wolf Creek, Stony Creek, Rye Hollow, etc., are dotted over with farms, even high up on the mountain sides. Many of these graze considerable herds of cattle, besides being in part devoted to cereal crops. Some of them are tobacco producing. The average capacity of corn land is 30 bushels to the acre, while a crop here and there will go to 60 bushels per acre.

Wheat rarely exceeds an average of 20 bushels per acre, though it often goes to 30 bushels and sometimes higher. Rye is one of the important crops of the county. Oats do well, rarely ever failing in any season to make a good crop. Plaster or gypsum is the fertilizer most commonly used, and is reported to increase the yie'd on nearly all the lands of the county, especially where there are rocks containing potash, soda, or lime.

The sulphuric acid contained in the gypsum, when set free, seems to be an easy solvent for a great many of the rocks showing on the surface in Giles County. A good deal of thin surface has the white flint, which contains potash and soda felspars derived from a vein of it about ten feet thick, which lies about 75 feet above the celebrated vein of iron ore. The dip of the rocks, as you approach the great basin of the county from any direction, is so gentle, that these rocks when found on the surface continue in sight for considerable distances. Gypsum when sown where even a small percentage of this rock is on the surface, increases the yield of the lands surprisingly. This may be owing also, somewhat, to small percentages of phosphates set free by the action of different reagents resulting from the decomposition of the gypsum. It is a good thing that, not only here but in many other localities, this cheap fertilizer, which abounds so plentifully in some of the counties of Southwest Virginia, should be so valuable and efficient.

FRUITS.

The apple seems to be the most successful and reliable of the fruit crops. It is an exception, rarely ever repeated twice in a century, that the apple crop fails. There are many improved kinds of apples in the county, the tree seeming to be long-lived and very fruitful. Peaches are very precarious. The trees seem to thrive well, but do not produce regularly. The pear and quince hit oftener. The cherry is a native, apparently, judging from its reliability and plentiful crop.

GRAPES.

This county has few but the native varieties, and they generally do very well annually. Not much attention, however, is paid to their culture.

BEE CULTURE

is slowly gaining ground in the county, and when readier means of transportation are supplied much honey will no doubt be raised for market.

MINERAL SPRINGS AND WATERING-PLACES.

The mineral springs of this county are not so numerous as those of Montgomery ; but at the New River White Sulphur, the water is said to be a highly medicinal and curative agent.

These springs are located on the east bank of New River, toward the south side of the county, about eleven or twelve miles north of the Atlantic, Mississippi and Ohio Railroad (New River Station). They will be right on the line of the New River Railroad when that road is completed, and will then enjoy that popularity to which their fine location and the curative power of the water entitles them.

Just here New River flows with a sweeping curve under one of its towering limestone cliffs, and, altogether, the surroundings go to make up one of those pictures of delightful mountain scenery which have as much influence in restoring the wasted energies of the human frame as any other cause.

Mountain Lake is a celebrated watering-place on the top of Salt Pond Mountain. Its success as one of the most renowned and frequented places of the kind, depends only upon facilities to get there. Its chief attraction, besides the pure mountain air and water, is the mountain lake—a sheet of water about three-fourths of a mile in length by a less width. It has not been in existence much more than a century. Perhaps the early settlers in salting their stock in its basin caused such a trampling of cattle over the small vents in the bottom of the basin that they were closed, and the lake thus began to form. Not many years since the forest which grew over this ground was visible below the water. Some of the old trunks are said to be still in view, as the water is very clear.

One of the features of that mountain after this, also, is to be the rearing of the Angora goat. A great many of them will soon be placed on these mountain sides, from which ample returns are expected in the wool they will produce.

SCENERY.

If some patent right could be secured by which all classes of fine scenery could be adequately described in a few words,

no doubt any writer on Giles would eagerly avail himself of
it : for no one can do justice to the subject here.

Angel's Rest, seemingly the guardian of the beautiful widespread emerald plain below it, is fitly named. Its beauty, as
it towers nearly 2,000 feet above the river below, and the fine
air which pours fresh down from its beautiful sides, would
seem to have the power to inspire any population living near
it with noble and elevated thoughts and feelings.

Some miles away, to the east, the Salt Pond and Butte
Mountains seem like great counterparts of Angel's Rest, and
lend an additional charm to the great scene in which New
River forms a distinct feature as it winds its way between
them.

Often repeated along this river are high cliffs, such as those
at the White Sulphur Springs, Clyburn's Ferry, Humphrey's
or French's Cliff, and numerous others. Rising sometimes
from 300 to 500 feet precipitously from the water, stained in
many colors of drab and red, brown and black, they present
a pleasing and attractive picture. There are pillars and
towers and columns, frequently suggesting the idea of design.
Nature has sculptured and decorated them in designs of more
than mortal conception of the beautiful.

Again, in the deep gorges of the high mountains, streams
like Mill Creek have poured their never-failing crystal waters,
unnoticed through time, over the high falls and cascades
almost concealed in the wealth of luxuriant vegetation of the
rich hollows.

Mountain Lake needs no eulogy! No description of it
could be rose-colored, viewed in sunshine or in storm. The
simple truth as to its history and natural beauty makes it
appear the creation of the highest fancy. Left alone it must
finally commend itself to any enlightened and appreciative
people.

FALLS AND CASCADES OF MILL CREEK, GILES CO., VA.

(P. 154.)

TRADE.

The trade of Giles is chiefly in cattle, horses and mules, sheep, wool, wheat, corn, and tobacco. Of the former, there are about 3,500 annually shipped—much too small a number for so good a grass county. Of this number, perhaps one third goes to the English markets. Of sheep, there are probably not more than 1,800 shipped, owing to the destruction occasioned by want of watchfulness on the part of owners. Of wool, there are about 10,000 pounds shipped annually; a small proportion of the wool grown is carded and spun at home.

Of wheat, there are 35,000 bushels now shipped. No doubt this quantity would be greatly increased with proper lines of transportation through the county. Of corn, there is but a small surplus shipped.

Tobacco is getting to be quite a staple in the county. Its product last year amounted to nearly 350,000 pounds. The average price received was from seven to eight dollars per hundred pounds, the lands rarely averaging more than 750 to 800 pounds per acre. Wolf Creek, and some points on lower Walker's Creek and New River, east of Pearisburg, seem to be the more prominent tobacco localities.

Any railroad line built through the county would soon cause an increase in the production of all staples. There is but little now to stimulate the population to raise more than will supply their wants and pay taxes.

MANUFACTURES.

There are no manufactures worthy of notice, beyond a few carding machines, though there is ample water-power on nearly all the streams to warrant extensive establishments.*

---

* Under this head may be described " The Sinking Creek Furnace," two miles east of Newport, on Sinking Creek. This furnace dates back to '73. It has had the usual experience of furnaces located eighteen or twenty miles

SCHOOLS.

There are two or more select schools in the county not connected with the public school system. The public schools have not given satisfaction in the last few years, but there seems now to be a chance that a better state of things will prevail in the future.

FISH CULTURE.

The attempts which have been made to stock New River with improved fish will no doubt show favorably in that stream and tributaries in the next few years. The black bass, which are becoming so numerous in the county of Pulaski, in New River, will soon fill the streams of Giles County.

TOWNS, ETC.

Pearisburg, containing the court-house, is situated in the shadow almost of the beautiful Angel's Rest. It contains churches of different denominations, schools, hotels, stores, cabinet-making establishments, smiths' shops, etc., etc.

Newport, toward the southeast side of the county, is a village pleasantly situated near the northern base of Gap Mountain. It has also schools, stores, and shops of different kinds. A town is springing up at the Narrows of New River, on the north side of the county, destined to be a great manufacturing place in iron and steel when all the railroads are built which are now contemplated to pass through it.

Staffordsville, Poplar Hill, and White Gate, on Walker's Creek, are notable places in the county.

---

from railway transportation. This furnace, with proper care observed in selecting its fluxes, will have a good future before it. It can select its ores from semi-magnetic or brown ores, as it chooses, and ought to make eight tons daily of prime charcoal pig. It is now in blast, being run by Mr. Brown, the banker, of Baltimore.

TRANSPORTATION LINES.

New River is contemplated to be made navigable at some day by the United States Government, surveys having been made to ascertain practicability, cost, etc. Some appropriations have been made and work done on the river, but none in this county.

The New River Railroad Company, under charters from the States of Virginia and West Virginia, is now constructing a narrow-gauge railroad along the banks of New River, in Giles—a road which will connect Hinton on the Chesapeake and Ohio Railroad with New River Station on the Atlantic, Mississippi and Ohio Railroad.

This road will be completed in two years, and will add much to the wealth of the county by the developments it will make of the vast iron deposits, the facilities it will afford, etc.

The line of the Virginia, Kentucky, and Ohio Railroad passes through the southern part of the county—a road which may be built in the next few years.

The line of the Shenandoah Valley Railroad extension also passes through Giles, following the line of Big Stony and Wolf Creeks.

The line of the Richmond and Southwestern Railroad passes also through the county from east to west, a narrow-gauge double-track road, leading from Richmond, Virginia, to Pound Gap in Kentucky, and on to the Mississippi River.

The line of the extension of the Richmond and Alleghany Railroad, now being constructed up James River, if built, would pass through this county from east to west.

The Pittsburgh Southern Railroad, now being built from Pittsburgh, Pa., southwardly, will, if extended into this region, pass through the county from north to south, following nearly the line of New River; possibly passing out southwestwardly through Shannon's Gap, and up Little Walker's Creek, in the direction of the great Cranberry magnetic ore beds of North Carolina.

BLAND COUNTY.

This is essentially a mountain county in every sense, more than one half being mountainous. Though comparatively new, it has many points of interest, besides holding within its irregular surface many of the most valuable mineral deposits in this section. It is not without fine grass lands, fine streams, and splendid scenery. Its timber areas are highly valuable, and its mineral waters rank among the most effective curative agents in the country. Its population is more than usually industrious and enterprising, and with the encouragement which increased facilities of transportation would give, its citizens would be among the most active in developing the fine resources of the county. And, what is not generally known, Bland holds at the south base of Brushy Mountain, a very respectable coal-field of its own.

HOW BOUNDED.

Bland has for its northern boundary line, the crest of East River Mountain; for a short distance is a border county, touching the county of Mercer, one of the most southern counties of West Virginia. It is bounded south by Wythe County and a part of Pulaski, east by Giles County, and west by Tazewell and Smyth Counties. The marked feature on its southern line is Walker's Mountain; on its western and northwestern, Garden and Round Mountains.

HOW WATERED.

Bland is finely watered by never-failing mountain streams of pure water, several of the prominent streams of the section taking their source in the county. Walker's Creek, which flows east into New River through Giles, rises in Bland and unites at Kimberling Church with Kimberling Creek, which has its source principally in Bland. The North Fork of Holston River rises near Sharon Springs and flows westwardly into Smyth County; and Wolf Creek, which flows out of Burk's Garden—receiving one of its affluents, Hunting Camp Creek, wholly a Bland County stream, at the end of Round Mountain—flows through Rocky Gap, where it unites the waters of Clear Fork and Laurel, on the northern side of the county, and enters New River near the Narrows in Giles County.

GEOLOGY.

The cross sections show Bland to differ somewhat from its neighbors in geological structure. Giles County would be very similar but for the coal measures of Bland, before alluded to. It has very much the same geology as Tazewell, except there is an apparent difference in the structure of the coal measures, as well as in the number of faults creating mountain ranges. In the short space of ten miles across the trend, Bland has no less than six, and a part of the way, seven of the most considerable mountain ranges in the State. It is in the very heart of the line of the Alleghany range prolonged southwestwardly, and as to the whole Appalachian system, including the Blue Ridge, Alleghany, and Cumberland mountains, occupies a central position. Bland also occupies an elevated position, giving rise to the waters of the Holston, which flow westwardly toward Tennessee, and the tributaries of New River flowing in the opposite direction.

The general elevation of the valleys is from 2,500 to 2,800 feet above sea level, while some of its mountains attain a height of 4,400 feet above the ocean.

A part of the southern border of the county overlapping
Walker's Mountain and taking in a portion of Little Walker's
Mountain's northern slope, it would be proper to say that the
southern end of the cross section begins about the Catskill
sandstone, which generally is about the central ledge of Little Walker's Mountain. Thence going north you pass over
the upturned edges of the underlying Chemung, or Old Red
Sandstone series, dipping 40˚ southwardly ; then the Hamilton slates and sandstone, generally thin bedded ; then the
Marcellus black slates with occasional crystals of lead and
copper, and sometimes so highly bituminous as to yield
nearly two inches of impure coal ; then still to the north,
with about the same dip (40˚) we reach the rocks in the
southern slopes of Walker's Mountain : first the Upper Helderberg about 40 feet of a flint ledge, sometimes giving a
silicious iron ore, and again lead and zinc sulphide in small
quantities ; then the Oriskany sandstone, sometimes 60 feet
thick, with 15 feet of its lower part so heavily charged with
iron and manganese, as to be a valuable ore bearing series ;
then the Lower Helderberg, which is sometimes so thin as
not to be noticed, while occasionally it assumes a thickness of
8 or 10 feet of limestone, more or less charged with lead sulphide. Then next to this north is a band of undetermined gray
sandstones of about 350 feet ; then the Clinton with 16 inches
of fossil ore ; then the Medina with its mottled sandstone
and heavy ledges of ironstone, perhaps 50 feet thick ; then
about the heart of Walker's Mountain outcropping at its
crest, is the Oneida sandstone not more than 40 feet thick ;
then, as you descend the northern slope, the calcareous sandstones and limes at the upper part of the Hudson River
group ; then the main body of the Hudson River 650 feet
thick ; then near the northern base, the Trenton limestones,
not more than 300 feet thick along here, with their base
marked by the felspathic flint measure, which seems to divide

it from the Upper Calciferous limestones; then the Upper
Calciferous, with some ledges not more than 50 feet below
the flint measures, marked by very flattering quantities of lead
and zinc sulphides; then the next limes of the Calciferous; but
a short distance leads us to a fault which brings us suddenly
against the rocks of the Proto-carboniferous, holding several
veins of coal, in the south spur of Brushy Mountain—one of
the veins measuring 8 feet, with varying dips between nearly
flat and nearly perpendicular, having a trend, like all these
rocks, about north 70 east. Then after this, going north over
400 or 500 feet of thin slates and sandstones, some of which
are conglomerate, we are again at the Catskill sandstones,
occupying the heart of Brushy Mountain; thence down its
northern slope, through Chemung and Hamilton to the Marcellus, in the eastern prolongation of Poor Valley. Again
here, at the foot of Round Mountain, on the south side, which
is a continuation east of Garden Mountain, are signs of lead
in the slate; then passing up on to Round Mountain, over
the Upper Helderberg flints, we are again soon in the Oriskany sandstone, which yields occasionally, along this mountain, a splendid deposit of iron ore of the character of that at
Chestnut Flat in Giles County, and sometimes a fine oxide of
manganese; next to this are the Lower Helderberg limes,
sometimes 75 feet thick, as in Garden Mountain, Round
Mountain, and Flat Top Mountain; then, in a few hundred
feet more, up Round Mountain, we again encounter the fossil
red ore about two feet thick, and then on the broad crest of
the mountain, the Oneida sandstone, assuming a rather anticlinal dip; then, as we descend on the northern side, a partial
repetition of the rocks on the southern side, dipping northwardly instead of southwardly as before. Round Mountain
has a trend northeast, but as you approach Rocky Gap, it
subsides and rises again in the Flat Top Mountain in the
eastern side of the county, next to Giles—the intervening

11

space being mostly marked by Devonian rocks characteristic of Brushy Mountain.

Then, pursuing the cross section line still north, we encounter an irregular synclinal fold at Wolf Creek, with the Marcellus slates, etc., very visible ; going up Rich Mountain side, on the south, we have a repetition of the south side of Round Mountain ; thence down its north side to Clear Fork, where we meet with a fault bringing the Trenton limestones into contact with Upper Silurian rocks, composing Buckhorn Mountain ; thence we pass on into East River Mountain, showing a repetition of the south of Round Mountain again.

IRON ORES.

The brown ores which Bland County would yield under the stimulus of cheap transportation, would be enormous in quantity, and many of them of a highly superior quality. The Oriskany measures on the south of Big Walker's Mountain, occasionally along its length in Bland, yield good brown ores in beds not over ten feet thick ; but the greater and more available masses of brown ores, toward the southern limit of the county, are in the Walker's Creek valley, about Newberry's and the line of the north side of Walker's Creek. These ores are due to the decomposition of pyritous veins near the felspathic flint, belonging about the junction of the Calciferous and Trenton limestones. It would be difficult to tell the number and thickness of these veins, as there are yet no developments of any consequence. Judging from the quantities on the surface in different places, they must be of good size, though no doubt variable in thickness. This series gives often, in different parts of the county, a semi-magnetic red ore, as well as true specular, together with the brown ore, being in the same zone identically with the great Giles County basin.

Brown ores occur in the coal rocks, but not in very large

quantities. The next great deposits of brown ores of any consequence are those in, and resulting from, the decomposition of the Oriskany measures in Garden Mountain, Round Mountain, Rich Mountain, Flat Top, and East River, and Buckhorn mountains. It has already been said that Round Mountain has an ore, on its south side, similar to that of Chestnut Flat in Giles. It may be as well to add that this particular ore bed in Round Mountain is near Kidd's Hunting Camp, 550 feet above water level, and is only second in size and importance to that of Chestnut Flat itself. It has about the following composition :

Sesquioxide of iron ....90.000 = 63 p. c. metallic iron.
Silica 2.500
Sulphuric acid 0.350
Phosphoric acid 0.280

This ore is regarded by experts as one of the best ores in Virginia. While it is classed with the brown ores, since it is found in the Oriskany measures, it really gives a blood-red when crushed, and has more the appearance of specular than the Chestnut Flat ore.

Brown ores are found in the Upper Helderberg series, mixed with a red ore in a vein about five feet thick, in Garden Mountain, Round, Rich, East River, and Flat Top mountains. The measure, though distinct at many points, is however obscure at others.

Brown ores again occur on Clear Fork, principally on the Buckhorn Mountain side of the stream, in flattering quantities, and of a quality superior for its easily fusible qualities, and general freedom from impurities.

Red Ores.

The two red ores of greatest quantity are the fossil red—in the larger mountains—and the fine specular which shows

now and then in the felspathic flint measures above mentioned, as in the hills toward Tillson's Mill, in the west end
of the county, and other points along that range, north bank
of Walker's Creek.

The fossil red is generally in one vein on the southern exposures, from sixteen inches to two feet thick, and would
yield for the whole county an unlimited tonnage. Walker's
Mountain, Round, Garden, Flat Top, Wolf Creek, Rich, and
East River mountains show the ore very distinctly, and it is
generally continuous. Some good judges, such as PROFESSOR
LESLEY, think it rather silicious, but it may be submitted that
as an ore to mix with the other abundant ores of the county,
it will serve a fine purpose.

The specular, or semi-magnetic red, of the Giles County
basin prolonged, is found near the felspathic flint in quantities, now and then, to be of great value; its great purity
and high percentage of metallic iron make it a very valuable
adjunct to the other ore deposits of the county.

Chromic ore has been reported from this county, but it is
questionable whether a ton of it will ever be found.

MANGANESE.

Ores of manganese are very abundant in this county.
Round Mountain—in the Oriskany measures—Flat Top, Garden, East River, and Rich mountains show veins of it sometimes over ten feet thick. Very frequently it is a very pure
binoxide. A little further exploration and development must
show it in such quantity as to make it quite an item of transportation.

It is scarcely necessary to give its analysis here. The
mass of it is not binoxide, but there is a great deal of it in
pure crystalline form, and would give the standard analysis
for pure ore whenever assayed.

COAL.

The coal of Bland, of any value, is in one vein, lying at the south foot of Brushy Mountain, and extending from the west end of the county to a point a few miles east of Seddon, where the coal measures have been uplifted and denuded.

This vein, owing to the disturbance of the whole formation, is occasionally ten feet thick—as near Sharon, for instance—but its general measure is 6 feet for the greater part of the distance, and sometimes not over 3½ feet. It yields, for parts of its length, a very firm bituminous coal; again, it gives a crushed article; but much of it is really valuable, though the dip is variable, between nearly vertical and horizontal. It may be said to measure from the outcrop down to the line of fault, or where it is cut off by limestone, over a half mile.

LEAD AND ZINC.

Lead and zinc are found in the most flattering quantities at several points, in a measure which underlies the felspathic flint lead; occurring all the way from Smyth County to the Giles County line; but on the turnpike 5 miles east of Sharon it shows more conspicuously than anywhere. The dolomite here dipping, first gently, then steeply to the south, has good lead and zinc sulphides shot through it in masses which sometimes weigh ten or twelve pounds; but as to whether it is there in a large and compact vein can only be determined by further development. It has a very flattering appearance on the outside; and some of the ores, both of lead and decomposed zinc, are of a very high percentage of purity. Again, in the Water Lime groups of the Lower Helderberg rocks, in Garden Mountain, as well as Flat Top Mountain on Dismal Creek, lead and zinc occur as very interesting constituents of the rocks.

BARYTES.

Barytes is now and then found near the felspathic flint lead also.

Copper is found in the Marcellus shales on either side of Round Mountain, but scarcely sufficient to pay for working.

Salt should naturally exist in the rocks of X. and XII., so abundant in Brushy Mountain.

Petroleum.—Though the petroleum rocks of Bland may be barren of oil, they are quite easily distinguished at the north base of Brushy Mountain. In the Kimberling district, about and below Kimberling Springs, there is a very large basin of converging dips, where the rocks ought naturally to form a reservoir for the oil drainage. This section is underlaid with the oil series at only a few hundred feet depth; and is really the only *apparent* oil basin outside of the coal measures, belonging strictly to the great Kanawha coal basin in this section.

BUILDING STONES.

The principal building stones are found among the limestones; also in the flags in large measures of thin-bedded sandstones of the Devonian rocks, underlying and north of the coal veins.

MINERAL SPRINGS.

Sharon Alum and Chalybeate Springs are situated on the turnpike leading from Wytheville to Tazewell, about 18 miles from the former, toward the western portion of the county. These springs, being arranged for the accommodation of numerous visitors, are justly regarded as among the most pleasant watering-places in the mountains. The elevation here is about 2,800 feet above the sea, and the fine water, combined with the pure air and healthy diet, make it an agreeable place to every one who visits these springs.

Kimberling Springs, in the central portion of the county, are noted for the wild and picturesque scenery surrounding them, the highly curative power of the sulphur water, and the invigorating influence of the fine air. These springs boast four different springs, one of which is an Alum Chalybeate. They are situated 28 miles north from Wytheville, Atlantic, Mississippi and Ohio Railroad.

TIMBER.

All of the timber enumerated for the other counties of Southwestern Virginia are represented in Bland County, except the balsam fir tree. There are large quantities of white pine in the Brushy Mountain, Hunting Camp Creek and Kimberling, toward the heads of its tributaries. White oak is abundant in nearly all sections of the county; large bodies of chestnut oak and chestnut. Walnut is common in the Walker's Creek and Holston Valleys and the Clear Fork Valley. Poplar is not now so plentiful as formerly; but deficiencies in other woods are made up by the quantity of white oak stave timber remaining in the Hunting Camp, Lick Creek, Kimberling County and other districts, and the great quantity of chestnut oak good for tan bark. There is also a great deal of hickory and the white woods, such as white walnut, cucumber, linn, buckeye, wahoo, and some ash, locust, and spruce pine.

WATER POWERS.

Numerous water powers are easily obtainable on Walker's Creek, Holston River, Wolf Creek, Clear Fork, Hunting Camp and Kimberling. These streams fall generally about 25 feet per mile, and will give powers requiring any measurement from 80 cubic feet per second down to 10 feet per second.

MANUFACTURES.

The people of this county with fairer opportunities would be decidedly a manufacturing population. To the extent of their ability, they are now very much disposed to combine for the purpose of erecting woolen mills and other factories.

There is now a good carding machine and woolen mill in the vicinity of Mechanicsburg, and others perhaps contemplated in the county. There are numerous good saw and grist mills in the county.

AGRICULTURE.

The grass fields of Walker's Creek and Holston Valley graze a great many cattle, sheep, etc., annually, and farming is carried on all over the county as extensively as the mountainous nature of the country will permit. Clear Fork, Wolf Creek, Hunting Camp, and Kimberling, also have good farming areas; and there are many good farms still in the forest uncleared. Sheep raising would be very profitable on a large scale in Bland; and for the Angora goat, no doubt it would be excellent.

SCENERY.

The scenery in Bland is in some parts imposing, on account of the greater field of view being taken up by mountains; but there are many choice views in the county. Frequently, a most lovely and romantic view will break upon the tourist or traveler in riding along the roads, which now and then lead through the mountain gorges, usually flanked on one side by a stream pouring over waterfalls and cascades, both road and stream fringed with rhododendrons, azaleas, and flowering shrubs, towering above which are apt to be thick spruce pines with their thick and dark foliage.

All the fruits of this latitude are apparently at home in Bland, the peach only being somewhat irregular in bearing.

Grapes are recognized as being peculiarly adapted to Bland ; and on Wolf Creek, on the north side of Round Mountain, a good deal of wine is made annually from the native varieties. The White Muscadine is one of the varieties native to Bland, and is said to be a well-flavored grape. *Bees* do well in Bland, where there are so many flowering shrubs and trees.

TRADE IN CATTLE, SHEEP, WOOL, WHEAT, AND CORN.

Bland annually sends off :

Cattle	2,850 head.
Sheep	4,000 "
Wool	11,000 pounds.
Wheat	2,800 bushels.

Corn. No attempt is made to raise corn for market, though a few hundred bushels are sometimes sold out of the county.

LINES OF TRANSPORTATION.

The lines of railway bidding fairest to be built at an early day through Bland, are the Richmond and Southwestern Railway and the Virginia, Kentucky, and Ohio Railway. Either of these roads would very fully develop the fine resources of the county.

TOWNS AND VILLAGES.

Seddon, the county site, is a place of 300 inhabitants, nearly at the middle of the county, east and west ; having besides the court house and the public records, good hotels, stores, churches, cabinet and smith shops, and an enterprising and progressive newspaper, called " *The South and West.*"

Mechanicsburg is somewhat smaller, in the southeastern part of the county, near Walker's Creek and Kimberling Church, with stores, etc.

Rocky Gap will one day be a manufacturing place, with its

fine water power, besides the gateway of great lines of railway.

Sharon has been mentioned in connection with Sharon
Springs.

PUBLIC SCHOOLS.

The public schools of Bland are now, like the other public
schools of this section, improving.

TAZEWELL COUNTY.

Of all the counties in Virginia, which may have justly
merited the praise bestowed upon them, none could receive
all the adulation which the utmost ingenuity could devise,
and still merit more, unless it were Tazewell County.

It looks as though some special attempt was being made to
give a rose coloring to all the subjects treated in this volume,
judging from the descriptions as they read ; but let the reader
once investigate for himself, and he will then see that the
powers of description and illustration are tame by the side
of the subjects treated.

Tazewell County, if in Europe, would be an empire within
itself. Its territory is considerable, being forty miles in
length by eighteen miles in width ; and, within that area, holds
a wealth of blue grass lands which are the admiration of all
who see them, both for their fertility and wide extent ;
splendid coal veins, lying well for mining ; iron ore deposits,
rich and extensive ; matchless mineral waters, and forests of
timber rarely, if ever, surpassed anywhere. Some attempt
will be made to describe its scenery further on ; but to do so
will be an ignominious failure ; for what pen could possibly
describe Burk's Garden and surroundings ; or, the view from
Dial Rock ; or, for the matter of that, the Cove and the grand
country about Liberty and Maiden Spring ; or, the mouth of

Indian, the river country, Bluestone and Wright's Valley? The day will surely come, when an appreciative traveling public will throng this county! Its citizens, who have been reared in the county and have become used to its every feature, are its enthusiastic admirers. They would not leave it hardly for any other spot on earth, so fully are they imbued with its loveliness, its fine water, pure air, and a noble future, heavy with the promise of a fine destiny soon to be fulfilled.

HOW BOUNDED.

Tazewell is one of the border counties, being bounded on the north by the county of McDowell in the State of West Virginia; east by the county of Mercer, West Virginia; southeast by Bland County, Virginia; south by Smyth County, Virginia, and west by Russell County. The northern boundary line follows generally the crest of Sandy Ridge, which, toward the northeast, takes the name of Great Flat Top Mountain; south, the great Clinch Range marks the boundary until you reach the east side of Burk's Garden, when it deflects to the northeast, crossing the Rich Mountain, and strikes East River Mountain near Nye's Cove; whence the line continues on the crest of the last-named mountain, eastwardly, to the State line.

HOW WATERED.

The greater part of the county is watered by Clinch River and its tributaries flowing southwestwardly. Bluestone River, East River, and Wolf Creek, with some of its tributaries, rise in the eastern portion of the county and flow eastwardly into New River. By this it is understood that Clinch River and some of the tributaries of New River have their source in this county, at a general elevation of about 2,900 feet above sea level. The whole county is well watered. Many of the springs send forth such a large supply of water as to be capable of running a grist-mill within a few hundred yards of their source.

GEOLOGY.

The geology, like that of Russell, is comprised between the rocks of the Lower Silurian and Carboniferous periods, including some of the Calciferous limestones below and a great part of the coal measures above.

Beginning on the south, in Clinch Mountain, Garden Mountain, Rich Mountain, and East River Mountain, you encounter the series holding the dyestone group, the Oriskany, etc., dipping southwardly, at angles rarely ever exceeding thirty-two degrees. Thence, going north, you see next, on the north of these mountains, the variously colored limes of the Hudson epoch; after which on the surface, still generally dipping south, the Trenton limestones; and then, near the line of Clinch River, the Calciferous limestones, a short distance north of which is the great fault that brings these lower rocks up into contact with the coal measures; thence, northwardly, for an average width of three miles, on the north side of the county, the coal and associated sand-rocks, slates, etc., form the chief feature. The accompanying cross section will present more graphically than it can be written the order of position of the different strata, and will serve, with but little variation, for almost any part of the county east or west. In the description of the different ores, etc., this cross section will be more fully explained.

IRON ORES.

To use an old comparison, it would be very much like the play of Hamlet with Hamlet left out, to attempt a description

Section through Western half of County.

Section through Eastern half of County.

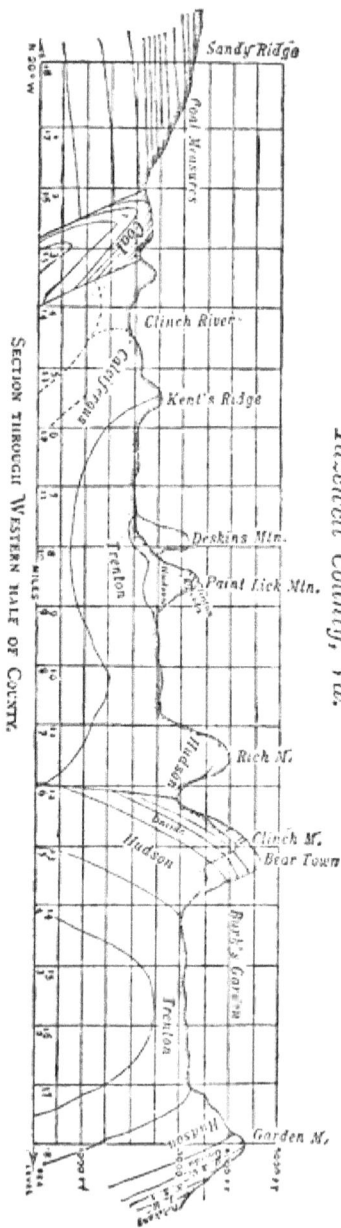

Tazewell County, Va.

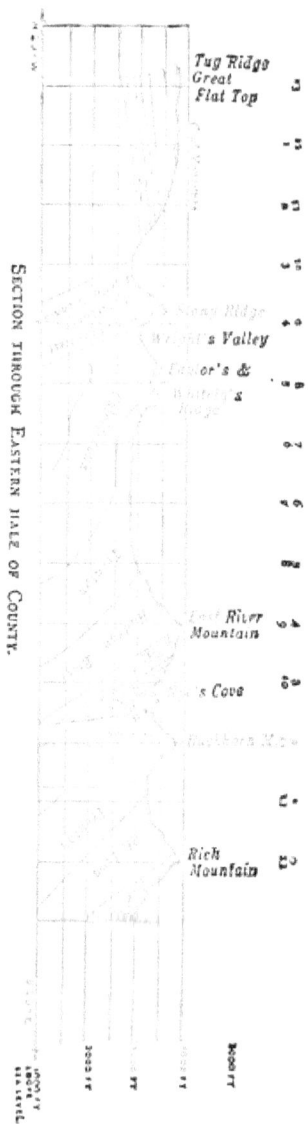

of Tazewell with the iron ores left out. The county will always be famous for its incomparable grass lands; but, when any line of transportation shall have been constructed competent to bring the ores and coal of Tazewell into communication with each other and with the other vast beds of ore in the country, it will then be seen that the iron ores are quite capable of performing the great task of bringing as much money into the county as the grass lands now do; and more, of helping other mining enterprises to stimulate farming and grazing to a much higher level of productiveness and profit to individuals and the county at large. It would be difficult to estimate the quantity of brown ores capable of being mined in the line of Kent's Ridge, Baptist Valley Ridge, Whitely's Ridge, and Taylor's Ridge, to say nothing of the admirable ores of Nye's Cove, Clear Fork, and other places. But to begin on the south side of the county, taking the various lines of deposits of brown ores, *seriatim*, through to the north side, will, perhaps, make the matter clearer. The first brown ores on the south side of consequence are those accompanied with manganese, which the Oriskany measures yield, on the south side of Clinch Mountain, bordering Poor Valley on the north, and lying in this county along where the four foot road to Marion crosses; thence east toward Bear Town. These measures will yield disconnected measures of a brown iron ore of high grade in masses sometimes fifteen feet thick by variable lengths, rarely ever exceeding 300 yards at one point; in this, following the habit of the Oriskany measures throughout this region. Then again, in the Upper Helderberg group, next overlying the Oriskany, a brown and red hematite, somewhat silicious, is found in a rather regular measure, about five feet thick, and continuous for considerable distances. This ore is usually very compact and hard, and is quite a new feature in the reading of the geology of this section, being the first time it has ever been brought before the public that this particular line of rocks is an iron-bearing formation. It is unfortunate that these ores are cut off from direct communication with the other ore beds of Tazewell by the Clinch and Garden Mountains intervening.

Both these lines of ores show their outcrops at an elevation generally of about 150 feet above the level of water in the small creeks near by.

In the south of Rich Mountain brown ores must show again in respectable bodies, derived from the same system of rocks. Near the junction of Cove Creek with Clear Fork, at the southeastern corner of the county, there are very fine, easily reducible brown ores in deposits resulting from the decomposition of the Oriskany iron stone in Buckhorn Mountain. Then up this Cove Creek, in Nye's Cove, very large boulders of this soft brown ore are scattered over the surface, to the left of the mouth of the cove as you enter. In this cove, to the right of its mouth, is Iron Ridge, an elevation of about 350 feet average, so named from the large quantity of brown ore showing on it. Then again on the spurs of Buckhorn and East River, in this cove, are other large bodies of brown ore. It may be submitted that fully 100,000 tons could be easily and cheaply obtained here at small cost for mining, without going to any appreciable depth after it.

At other points along Buckhorn and East River Mountains, west of Nye's Cove, the Oriskany and Helderberg series are the parent iron stores from which are derived bodies of excellent ore. The south spurs of Paint Lick Mountain likewise show deposits of brown ore, but generally mixed with red ore, due to the decomposition of the dyestone series, which occupies a great part of the crest of the mountain.

The next line of brown ore of consequence is found in the well-marked felspathic flint lead which outcrops in a line generally parallel with, and close to, Clinch River, known as the ores of Kent's Ridge, Whitely's Ridge, Taylor's Ridge, and Wright's Valley Ridge. All of these ridges are in a general line toward the north side of the county, a mile or so south of the great fault on the border of the coal measures ; Kent's Ridge lying westwardly, and Whitely's and Taylor's Ridges being toward the eastern end of the county. Elevation about 300 feet.

This series of rocks lies about the junction of the Black River and Trenton limes, and, in all probability, marks the closing of one epoch and the beginning of the other. The ore vein is composed of brown iron ore with oxide of manganese, in the proportion of about three of the former to two of the latter. At a few points here and there, both in Tazewell and Russell Counties, the ores of these ridges have been used with satisfaction in several forges now out of blast.

After this, going north, near the Mouth of Indian, and at other points along the margin of the coal fields, there are brown ores showing in considerable quantities, which may be ascribed to the decomposition of carbonates. It is to be regretted that so few developments of all these ore deposits should have been made ; sufficient evidence is given by the quantities exposed on the surface to prove that the vast amount of material which has decomposed in past ages has left behind immense beds of ore, besides what still remains in the original veins below the zone of decomposition.

Red Iron Ores.

As before stated there is red ore in the Upper Helderberg flint groups on the south of Clinch Mountain. Higher up on this mountain, on the south, is the band in dyestone red iron ore in the Clinton group, about eighteen inches or two feet thick, generally a silicious ore, but valuable as a mixing ore. This vein is generally continuous in this part of the mountain, and is about 800 feet above water at its outcrop. Further east, in Rich Mountain, in Buckhorn and East River Mountains, particularly about Nye's Cove, the dyestone group shows in large sections and fragments.

In Paint Lick Mountain, about its crest, it constitutes the covering of the great cliff-like wall of Oneida sandstone, which runs for seven or eight miles along near the top of this mountain, separated from it by sixty feet of intervening Medina sandstone and a few ledges of Clinton Red sandstone. The vein here is sixteen inches thick, and is only the lower of the dyestone veins. Its value will be chiefly as an ore to mix with other ores. Its quantity may be taken at an average width of 200 feet by about five miles, as it does not form a continuous belt, but is sometimes entirely denuded. It may be regarded as the parent of much of the ore appearing in the deposits in the valleys below.

North of this line there are no appreciable quantities of red ore in the county, except that occasionally the ores in the felspathic flint, above-mentioned, sometimes assume the form of a fine specular ore.

Iron Pyrites.

It is not yet determined that the source of the brown ores in the Oriskany Rocks is pyrites, but it may be assumed that such is the case with ore in the Upper Helderberg rocks. In the line of the ores in the felspathic flint—mentioned as being about the division between the Calciferous and Trenton limestones, there are, no doubt, below water level, very considerable quantities of pyrites. In the coal measures there are occasional pieces found, not generally of large size.

In the eastern part of the county, near the Mercer boundary line, there are ores reported of a slaty structure.

12

MANGANESE.

Manganese is very generally distributed through the county. There are some beautiful fragments of binoxide occasionally found, as in Whitely's Ridge, Taylor's Ridge, Kent's Ridge, Clinch Mountain, Nye's Cove, Buckhorn, Yosts, etc.; but no developments of it have as yet been made to an extent sufficient to give a very fair idea of quantity. Judging from surface indications there must be a great deal of it.

COAL.

The examinations made of the coal veins of Tazewell, on Middle Creek, Horse Pen Cove, and Abb's Valley, very fully and amply sustain all the declarations made by Prof. J. P. Leslie and others as to the size of the veins, the quantity they will yield, and general commercial importance. In fact, it may safely be insisted upon that these veins are among the most valuable of their kind in the great basin to which they belong; not only on account of the thickness of some of them, but the cheapness with which they can be mined and utilized, once the question of transportation is settled.

Beginning with Middle Creek, an affluent of Clinch River, just below the mouth of Indian Creek, it is there that some of the most reliable readings can be obtained, both as to that broad fragment of the measures inclining at an angle of 39° southwardly, and the area close at hand, to the north, giving the almost horizontal measures; the last continuing, with no great observable interruption, to the Ohio River.

Going from the mouth of the creek up about one and three quarter miles, over a fragment of Devonian rocks brought up by the thrust of Calciferous limestones against that side of the great fault, and over the limestones upon which this fragment rides, you encounter the sandstones and slates of what are presumably the Sub-Carboniferous measures, holding eight or nine veins of coal of different dimensions, generally dipping 39° south 20° east.

The first seven encountered have not yet been opened; but the eighth, being a large vein of good bituminous coal, very suitable for use in smith shops, has been mined. This vein lies between head and foot-walls of slate, with 50 inches of good coal next head-wall, then 12 inches of slate with 9 inches of coal next foot-wall; under this 15 feet of slate; then 54 inches of coal, of same dip, underlaid with slate; then, at about two and a half miles in an air line from the mouth of the creek, the horizontal measures set in which continue on for many miles.

These veins, to which the one measured belongs, may be said to continue for a great many miles in either direction. In this immediate vicinity they will give an average breasting, at a dip of 39°, of about 380 feet on the incline above the creeks, which cut through them about every two and a half miles. It is so easy for any one to calculate from this data the probable yield of this one vein above water, that it is needless to give it here. In the horizontal measures, this part of the county shows one vein about 80 feet below the crest of Sandy Ridge, which measures 4 feet clear coal, and without doubt contains a half-dozen other good workable veins, although there are no developments as yet.

Horse Pen Cove, situated about eight miles northwardly from Jeffersonville, the county site, is reported to contain not only an 8-feet vein of good bituminous coal, but cannel coal, as well as at Middle Creek. The cannel coal, though good, has not been found in a vein over 3 feet thick.

Abb's Valley presents no inclined dips of any consequence in the coal measures. They are almost without exception nearly flat. Near Smith's Store 4 feet veins are found close to the limestone of the valley. At the lower end of the valley, near the boundary line of the county on the northeast side, a remarkable vein of good bituminous coal, 11 feet thick, with a parting of slate, 1 foot, is easily accessible. This vein seems to lie at the base of the series just here, and is supposed to be the same one which shows near water level on Dump's Creek, in Russell County, there recorded 9 feet 4 inches.

Enough has been said to show that great quantities of coal and iron exist in Tazewell. Forty miles length of each, by variable thicknesses and widths, will present to the mind of the reader an idea of vastness, without the necessity to go into minute calculations to prove the assertion that the county will one day be as much noted for its mining and manufacturing, as it is now for its incomparable grass fields.

LEAD AND ZINC.

Small pieces of lead and zinc ores are occasionally found in the Kent's Ridge line of iron and manganese measures, and near the end of Taylor's Ridge, in the eastern end of the county, about twelve or thirteen miles from Jeffersonville, and two miles east of Springville. This lead ore really comes from a stratum of dolomite, lying apparently about 200 feet below the felspathic flint lead, which marks the iron and manganese above mentioned. In Nye's Cove, and occasionally at other places, the Lower Helderberg limes show in thicknesses varying between 6 or 8 feet and 75 feet. These rocks hold small quantities of lead and zinc ores, but are not supposed to exist in sufficient quantities to pay for working. Also in the Marcellus black shales and slates small crystals of lead ore now and then show. These measures are observable in only a few places in Tazewell County—Nye's Cove, and a part of the south of Clinch Mountain, west of Burk's Garden.

BARYTES.

Baryta seems to exist in the county in sufficient quantities to satisfy all the demands of trade in that article. A fifteen feet vein runs on the north side of Clinch River through the Cavitt's Creek country, and down through the Baptist Valley ridges all the way into Russell County, where it shows on the north side of Clinch River. This measure is between walls of flint on one side and dolomitic limestone on the other, in places; but it varies in thickness and quality very much. Where it is easily observable—as at a point in the main road leading from Jeffersonville toward Baptist Valley, on the north side of Clinch River—it shows well for thickness and quality. It appears to occupy a position about five hundred feet above the felspathic flint lead above mentioned.

COPPER.

Copper ore is sometimes found in traces in the felspathic flint, which it may be as well to repeat, is the stratum which lies near the division between the Calciferous and Trenton limes—carrying the iron and manganese ores.

SALT.

Falling Waters, toward the northwestern border of the county, seems to offer a true salt basin. The salt-bearing series underlies this part of the valley, and would no doubt yield brine if properly hunted after. A few miles lower down the Clinch River not only salt but petroleum seems to have been discovered oozing from the surface, but as yet nothing has been done of a satisfactory nature toward the development of the field.

BUILDING STONES.

Numerous ledges of both lime and sandstone exist in various parts of the county very well suited to building purposes.

Some of the limes are even ornamental, as those lying in such masses in the hills north of and about the court-house; no doubt the same gray masses which are near the division between the Trenton and Hudson epochs.

The sandstones, fit for building purposes, are those abounding in the coal measures, many masses of which are soft when quarried, but become very hard on exposure.

SOAPSTONE

Is found in ledges quite frequently as highly magnesian limestone, but no true steatite is found. Occasionally near the coal veins impure silicate of magnesia shows itself in thin beds.

MARBLE.

The large beds of variegated and gray limes, which have such a character as fine marble in Russell and Scott counties farther west, seem to have lost their distinctive character as such in Tazewell, and are not in such masses. In this it is not asserted that there is no marble in the county, for the strata about the base of the Hudson epoch, near the northern bases of the larger mountains, must show occasional beds.

MINERAL SPRINGS.

The Tazewell Springs, a sulphur spring situated five miles northwest from Jeffersonville near the Sandy turnpike, is a white sulphur water, containing such constituents as to render it diaphoretic in its effects. It is a good spring of its kind, and was at one time, before the era of railroads, a place of resort. Now that a railroad is likely soon to be constructed near it, will command attention again.

At Mouth of Indian, or Cedar Bluffs, there is a fine white sulphur spring on the bank of Clinch River, in a lovely and romantic situation. The water is charged with sulphate of magnesia and other valuable constituents, and flows from a series of magnesian rocks not far from the line of fissure separating the limestones from the coal measures. This spring has been provided with a neat basin hollowed out of a single stone ; and its fine curative powers, combined with its very convenient location, will make it a place of resort when a railway is built near it, and is even now much visited.

At Mustard's, in the eastern middle portion of the county near some of the head springs of Clinch River, are eight fine mineral springs very close together. Two of these come up on two sides of one thin ledge of limestone, one apparently is an arsenical spring, the other a blue sulphur, while the character of the others has not been determined. These springs are regarded as highly curative by those who have tried them ; they yield a fine supply of water. In many other places in the county mineral springs break out, but the above mentioned are those which command the confidence of those who have tried them as curative agents.

TIMBER.

This subject is difficult to treat with justice in a few words. The widespread fertility of the soil in Tazewell gave the county at one time a very large proportion of walnut, maple, poplar, linn, cucumber, etc. It is true that immense quantities of these valuable trees have been destroyed in clearing the lands, but that very large quantities still remain is equally true. Every tree that this latitude is capable of producing seems to grow to its utmost limit of size and beauty in Tazewell County, among which may be mentioned, white and black walnut, poplar, cherry, linn, cucumber, wahoo, hickory, ash, white oak, chestnut oak, red oak, black oak, pin oak, and other varieties of oak, maple sugar tree, or large maple, buckeye, mountain mahogany, birch, beech, dogwood, locust, elm, sycamore, yellow pine, black pine, white pine, cedar, hemlock, or spruce pine, and upon some of the higher points, such as Bear Town, extensive forests of balsam fir. Among the shrubs and flowering plants there are many of the most beautiful. In the spring the wide-spreading woodlands are resonant with the songs of birds of varied plumage, not yet invaded by that murderous nuisance the English sparrow.

To attempt to fix a commercial value upon the woods of Tazewell, would be idle. That can only be determined by the facilities which any line of railway running through the county may present. The walnut is in sufficient quantity to be the source of a splendid revenue. Such is the case with many other trees. Sugar tree, mountain mahogany, and other ornamental woods, besides white oak, will supply a large demand. In many localities, such as Nye's Cove, the chestnut oak will yield immensely in tan-bark. From this last source the county will derive a very considerable revenue.

WATER POWER.

The average fall of the larger streams of the county being rather over than under twenty-five feet per mile, it may safely be assumed that there is abundance of water power in the county. Clinch River discharges near the Mouth of Indian about 185 cubic feet per second, and has sufficient fall there to give three very good powers within two miles. Thence all the way up this stream to near its source on the main branch and tributaries, it is used to run grist and saw-mills, carding machines, woolen factories, etc. Maiden Spring Fork, scarcely inferior in size to the Clinch River, offers many mill sites, its fall being steeper. Blue Stone River, in the eastern section of the county, and the Clear Fork of Wolf Creek, are somewhat smaller than Clinch

River at Mouth of Indian, but they supply numerous water powers, only about one half of which are in use. Cove Creek, the stream which comes out of Nye's Cove, offers very good water power. Many small streams offer limited powers which would be useful for grinding. The never-failing character of these streams is one of their chief recommendations. A power calculated upon their ordinary discharge may be relied upon.

MANUFACTURES.

Hitherto manufactures have not been very carefully fostered in this community, but of recent years some very successful efforts have been made in that direction. The woolen mill, two and a half miles east of Jeffersonville on that branch of Clinch River, annually cards and spins a great deal of the wool of that vicinity and neighboring counties, and turns out a very creditable line of goods from its looms. Efforts are being made to increase the capacity of the woolen mills just above the Mouth of Indian, which have so long been in operation there. The locality is one that invites enterprise and capital. At this place is situated a tobacco factory having a capacity of about 150,000 pounds of manufactured tobacco. Both it and the fine furniture factory, located at the same place, will be greatly stimulated by the building of a railroad through that section.

AGRICULTURE.

In speaking of this subject it is difficult to avoid the use of effusive language. With the exception of a part of the coal belt, perhaps three fourths of its area is capable of producing fine grass, notwithstanding the county is traversed in its greatest length by more than one of the highest mountains in the Appalachian Chain. Even upon the very summits of some of these lofty ridges there are extensive grass fields, upon

which a large number of cattle, sheep, horses, etc., graze
annually.

Nature may have endowed other portions of the country
equally with Tazewell in the distribution of her choice gifts;
but it is doubtful if any other locality unites so many advantages adapted to grazing purposes as may be found here.
Not only is the grass a natural product, and of a highly nutritious kind, but the fields are nearly all without exception
well watered, and the small insects, flies, mosquitoes, etc.,
which annoy cattle so much at other places, are either absent,
or have so short a season in which to ply their vocation that
they are not injurious. With the least economy and care these
truly magnificent grass fields would sustain a fine population
in greater independence, ease, and even luxury than any other
section except the nearly similar districts of neighboring
counties. Throughout nearly all the valleys and far up, even
to the crests on some of the mountains, the highly fossil-bearing limes of the Trenton and Hudson epochs are adding
annually a new supply of fertilizing material to the soil,
decomposing to some extent with every rain in summer, and
more with each freeze and thaw of winter. The fossil shells
of these rocks are easily detected in nearly every ledge; and,
being charged to some extent with phosphate of lime, it is
easy to understand how their decomposition will affect the
soil beneficially. It is hardly necessary to cite localities in
which these results are prominently brought out. To use a
common expression, they are all over. The Cove, Burk's
Garden, about Liberty, Maiden Spring, the river two miles
east of Jeffersonville, Wright's Valley, Abb's Valley, Blue
Stone, Clear Fork, localities about Baptist Valley, the river,
the slopes of the mountains—and where not?

Not only is the land productive of the fine grasses, such
as blue grass—which is natural—with Randal grass, clover,
timothy, orchard grass, herd's grass and the like; but corn,

wheat, rye, oats, barley, buckwheat, and all vegetables yield largely in the rich soil.

It is not too much to say that Tazewell stands almost alone in Virginia for unaided natural capacity as an agricultural county.

SCENERY.

If attractive scenery can lend anything to the interest which might be induced by the other valuable features of the county, Tazewell could easily supply all that would be needed for the purpose, though the ability to paint or describe it might be wanting. Burk's Garden is an emerald sea in the spring time, with waving trees and noble pastures. 3,200 feet is its elevation above the ocean, with its border encircled by the Clinch Range of mountains, some peaks of which—as Bear Town—attain an elevation of 4,700 feet. Burk's Garden, though often described, it may be repeated, is about eight miles long from northeast to southwest, and about four and a half miles wide. It looks as though it had once been a mountain lake, the waters of which had burst their way through the northern escarpment that helped to hold it, leaving the beautiful trout stream that now pours through the gorge to mark its course. Geologically it rests in the center of one of the only great anticlinals in this section of the country. The great mountain, containing the Upper and Lower Silurian rocks, having been partially folded by a great pressure from the southeast, seems to have been parted along the crest of the anticlinal fold by a pressure subsequently exerted from the opposite direction, causing it to spread apart, leaving the limestones toward the base interjected between each other, so as to form the great plain now known as the great grass-producing Burk's Garden. There is no telling how much of it has been denuded and washed away in the course of time. Surrounding mountains and all must at one time have occupied a much higher position than they now do.

Thompson's and Ward's Coves form the wings of a great opening in the great mountain belt, somewhat similar in character to Burk's Garden, and offer, together with the country about Maiden Spring, Liberty, and Paint Lick Mountain, with its high, wall-like cliffs painted by the Indians, a piece of landscape scenery of marvelous beauty.

From Dial Rock, which is near the western terminus of East River Mountain, all that vast plain, with its grassy fields and wooded hills, about Jeffersonville, and the country east and north, look like the realization of Eden. From this point, the mountains seem to be marshaled rank after rank, as far as the eye can see. In some lights, the beholder is almost persuaded the great silent scene will take motion and march away.

Space forbids further attempt to eulogize in tame language a subject so perfect from the hand of nature. To see it only can it be realized.

The game and fish of the county are interesting subjects. A few bears still roam in the high woods and thick laurel of Bear Town Mountain, and the bass in the streams furnish fine sport.

The *Fruits* of the county are all those which this latitude produces so abundantly, and Tazewell fares like the rest in being very certain of a good apple crop annually, with the peaches only somewhat doubtful.

Grape culture has been very carefully attended to by several prominent gentlemen at the court-house and two or three miles east of there. In fact, these gentlemen have clearly demonstrated the high perfection to which the native varieties can be brought by proper training and attention.

Bee culture is engaging the attention of the people more and more every year, and with the proper stimulus in the way of transportation, would eventually become a paying industry.

TRADE IN CATTLE, SHEEP, WOOL, WHEAT, CORN, AND TOBACCO.

Cattle annually sold from the county....11,500 head

(Of this number a proportion, which increases annually, goes to Europe.)

Sheep.................................10,000 head
Wool, which is more than half used at
   home...............................30,000 pounds
Wheat..............................65,000 bushels
Corn......... 8,000   "
Tobacco............................10,000 pounds

There being no transportation now to distant markets, very little more wheat and corn are raised than will supply the home demand.

The trade in horses and mules is not very considerable, hardly enough to merit a notice. There is a decided spirit of improvement gaining ground in the minds of the people with reference to all classes of stock.

TOWNS AND VILLAGES.

Jeffersonville, the county site, is the chief place. It contains good schools, churches of different denominations, stores well supplied with everything generally sold in the better districts of country, hotels, and shops of various kinds for the repair of wagons, plows, smithing, etc.

Liberty is a village nine miles southwest from Jeffersonville, supplied with a hotel, stores, etc.

Mouth of Indian, or Cedar Bluff, at the junction of Indian Creek with Clinch River, is sixteen miles west of the court-house, a thriving place, having a hotel, stores, furniture and tobacco factories, saw and grist-mills, and will become a manufacturing center, being within a mile or two of the great coal deposits on one side and the iron ore on the other.

Springville and *Fall's Mills* are in the eastern part of the county, and they, with several other places of that size, form good trading places for their respective neighborhoods.

PUBLIC SCHOOLS.

Tazewell has always been careful to keep up its public schools. The appearance of the school-houses attests this fact, and now her schools seem to be more prosperous than they were some years back.

RUSSELL COUNTY.

This large county is noted for its fine grass lands, its coal, timber, and marble. It is not without large and valuable iron ore deposits; but its high character for extensive areas of fine blue-grass lands, thick coal veins, and heavily timbered forests composed of valuable woods, give it a name for both beauty and fertility, as an agricultural, mineral, and timbered county, shared by only a few of the other counties in the State. Could any landscape exceed in beauty Rosedale and Elk Garden? They are only equaled by a few places in Southwestern Virginia. In Russell, they may find almost a counterpart in Cassel's Woods, some parts of the county toward Hansonville and Dickinsonville, and occasional spots about New Garden, including the river scenery afforded by the different branches of Clinch River. It is singular that so much natural wealth should have lain so long among so progressive a people without fuller development.

HOW BOUNDED.

Russell is about 34 miles long from east to west, and 18 miles wide from north to south. The broad dividing ridge

which separates the waters of Clinch River from those of
Sandy River marks its northern boundary line, the next
county north being Buchanan County. South, Russell is
divided from Washington County by Clinch Mountain, a great
barrier which runs a course about north 70° east for many
miles through this section of Virginia. West it is bounded
by Scott County, and east by Tazewell.

HOW WATERED.

The county is well watered, mainly by Clinch River and its
tributaries occupying the northern half of the county. Its
southwest section has a considerable length of Moccasin
Creek, a tributary of Holston River.

GEOLOGY OF RUSSELL.

Russell, on the south, shows the rocks of the Upper Silurian age dipping southwardly. As you proceed north you
pass over the edges of the next lower strata, through the
Hudson River and part of the Trenton limestones, etc., until
you reach the great fault, north of the line of Clinch River,
which marks the southern limit of the great Kanawha Coal
Basin. Then for about five miles you are in the coal rocks,
carrying you to the northern boundary line.

Nearly all the strata of rocks pass through the county from
northeast to southwest, in a direction parallel with the course
of the mountain ranges; and hence a description of a section
taken at nearly any north and south line will have its material points very much the same.

The Clinch Mountain Range, having been subjected to some
cross flexures and end compressions, is duplicated more than
once in Russell; hence the iron ores, which are common to
the rocks of which it is composed, are brought out in considerable quantities. Thus, the southern border of the

county is marked by the dyestone and Oriskany series—
mainly on the mountain crests—giving valuable deposits at
intervals along the south boundary of the dyestone or fossil
red ore of very high grade, and of the brown ores of the
Oriskany. In the latter may also be found respectable quantities of manganese. Very frequently the Hudson River
limestones form the crests of these southern hills and mountains—as in the case of Rich Mountain—which, by the
decomposition of their highly fossiliferous strata, make a
rich and enduring soil, well adapted to grass, which is of
spontaneous growth. Passing north a few hundred yards to
the northern foot of the Clinch Range, you meet with the
thick ledges of variegated stone about the division between
the Trenton and Hudson series, looking now and then as if it
would make a fine marble. Close under this, outcropping to
the north, is a thick band of gray and pink marble, having a
fine texture, and admitting of a high polish. Passing northwardly, over Trenton limestones dipping in various directions, you reach Copper Ridge, which is a continuation
southwestwardly of Paint Lick and House and Barn Mountains, with the iron ore formation of No. 4 generally denuded.
The lower rocks of the Trenton in this ridge, holding the
ledges of felspathic material, seem to be brought up to the
surface, occasionally showing good deposits of iron ores and
fine crystals of manganese. In a short distance northwardly
you reach Clinch River, close to which, on the north side, is
the great fault, which brings an upthrow of Lower Silurian
limestone in contact with the coal rocks. Between the river
and this fault is a range of flint measures extending northeast and southwest through the county, which, for the greater
part, yields about 15 feet thickness of fine barytes, occasionally showing fine crystals of lead and a trace of carbonate of
copper.

    The rocks along here generally dip southwardly.   1

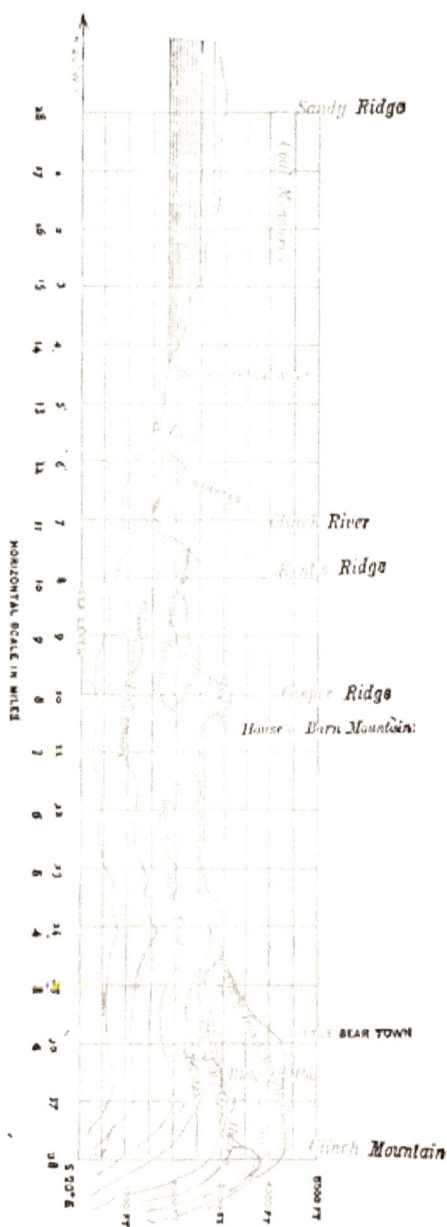

Section through Russell County, Va.

Sandy Ridge

Clinch River

Eagle Ridge

Copper Ridge

House & Barn Mountain

BEAR TOWN

Clinch Mountain

HORIZONTAL SCALE IN MILES

13

fault alluded to is frequently a double one ; that is, between the limestones and the horizontal coal rocks there is a great fragment of the coal strata dipping southwardly at a high angle, separated from the more regular measures on the north by a distinct line.

After this, until you reach the northern boundary line, mainly on the crest of Sandy Ridge, you are in regular measures of nearly horizontal lying strata of coal rocks.

IRON ORES

It would be very difficult to form an estimate of the probable quantities of different ores in the county of Russell. *The fossil ore*, of a very high grade, is about two feet thick, in several small veins, in discontinuous measures, in Clinch Mountain. *Brown iron ores* of the Oriskany, in the same mountain, are subject to the same conditions as the fossil ores, and so are the *manganese ores*. The *brown ores* in Kent's Ridge and the river hills, along down the line of the Clinch, are due to the decomposition of a vein of sulphuret, which is found throughout this tier of counties, about the junction of the Black River and Trenton limestones ; also on Cedar Creek, two and a half miles east of Lebanon, near an old forge site. This line of ores will prove to be an interesting adjunct to any railway line located near Clinch River. The thickness may be placed at six feet. The House and Barn Mountain, on the east side of the county, has unfortunately been long denuded of the masses of dyestone ore which once crowned it ; and it now only shows a deposit here and there, sufficient to delude the unwary into the belief that there are still large quantities of ore about the mountain.

It is possible that the ores now and then showing in Copper Ridge are the same in character and origin as those in Kent's Ridge and the river hills.

COAL.

There is nothing left to conjecture concerning the coal in this county, except, perhaps, its exact geological age. Where the veins are known to be so continuous, with good measures and occupying such an area, their position in proper geological succession is not of so much consequence. It may safely be assumed that a great part of the Lower Measures is here represented. The area in this county is about 100 square miles. The veins are thick and some of them are of cannel coal.

Up the Hurricane Fork of Dumps Creek, the lowest vein in sight—at the bed of the creek, near Grisell's—measured nine feet four inches, pitch one in twenty, south 80 east, with head-wall of slate fifteen feet thick, overlaid with thin bedded sandstones ; foot-wall eight inches of clay slate, underlaid by heavy sandstones of rather irregular formation. This coal is of the flaming bituminous variety.

One hundred and fifty feet above this is a vein of very much the same kind of coal, nearly five feet thick, cropping out at Rasnicks. One hundred feet above this is another vein of good coal, not yet measured at that point for want of proper openings. Fifty feet over this is a vein of cannel coal eleven inches thick, underlaid by quartzose, blackish sandstone, underlaid again by a heavy band of sandstone; the roof is three and a half feet of slate, overlaid by about 100 feet of sandstone. Over this is a series of sandstones and slates, with veins of coal, running up to an elevation above the creek of about 1,000 feet.

It was impossible to get at the exact number and thickness of the veins in the series here. No developments of the veins outcropping any distance above the creeks have been made ; but, that the coal exists in good veins, at intervals, all the way to the tops of the hills is a fact revealed here and there by land-slides in the steeper hollows and wet places.

The reading obtained on Dumps Creek, with some variations in the measures, will give a fair idea of the whole.

Mill Creek, Sword's Creek, the upper end of Lewis Creek, Thompson's Creek, Weaver's Creek, and Lick Creek, as well as Dumps Creek, rise in the coal measures and cut deep sections as they pass through the coal rocks on their way to Clinch River.

LEAD AND ZINC.

There is an outcropping of baryta-bearing rocks on the north side of Clinch River, not far below the junction of Maiden Spring Fork, which also shows some lead sulphuret; but it is not probable that any workable veins will be discovered. Zinc is not found in any quantity.

BARYTES.

Barytes, in a vein about fifteen feet thick, is found in the series of rocks about the upper part of the Calciferous rocks, marked by beds of felspathic flint. The quantity of barytes in the county, along the line of hills bordering Clinch River on the north, must be enormous. So far as observed it must afford an immense tonnage to any railway line that might be built near it.

COPPER ORE.

Small particles of carbonate of copper are found in the same rocks.

SALT.

The salt-bearing series of Nos. 10 and 12 underlie the coal measures in this county, at accessible depths.

LIMESTONE.

The limestones of Russell are chiefly of the Trenton and Hudson River series. They are generally very fossiliferous, many ledges affording excellent lime. About Clinch River the upper measures of Calciferous limestones show occasionally; weathering sometimes very light drab as if they were highly magnesian.

BUILDING STONES.

The coal measures afford a sandstone easily worked in the quarry, which, upon exposure, becomes very hard. Many ledges of limestone are suitable for building purposes.

MARBLE.

Perhaps no other mention of the marble may be considered necessary, further than to say that its great quantity toward the south side of the county is such as to render it an item of consequence to any railway line passing through the county.

MINERAL SPRINGS.

While there are many mineral springs of value in the county, no doubt, there are none as yet improved.

TIMBER.

It would astonish any one from the great timber and lumber markets, where everything of the kind has its value, to see the quantities of fine trees destroyed annually in clearing lands. Very frequently the log heaps are composed of walnut, hickory, white oak, and poplar, ready for the torch. It is the only way the people have of getting the timber that they do not want for fencing purposes out of the way. This county is supplied with vast quantities of fine timber—about the river the black walnut and the sycamore. In fact, there are few localities in the county where the walnut is not found. Occasionally, as at Little Bear Town, on a loup of the Clinch Mountain Range, the balsam fir tree is found in

considerable quantities. White oak abounds everywhere.
Poplar, cherry, locust, chestnut, chestnut oak, and other
varieties of oak, linn, sugar, buckeye, and several varieties
of smaller growth. In the hills of the coal measures nearly
all these trees, including the hemlock, abound in their
primitive size and beauty, and in quantities seemingly sufficient to defy extermination.

WATER POWER.

Clinch River, after the junction of the two principal
branches in the east end of the county, discharges about 220
cubic feet per second, and constantly increases its volume as
it descends. Its fall is sufficiently great to afford numerous
valuable powers. Cedar Creek, one of its principal tributaries flowing near Lebanon, the court-house, having a
rather high average fall per mile, offers a good many mill
sites. Moccasin Creek gives good power. Lick Creek, Indian Creek, Lewis Creek, and the other creeks on the north
side of Clinch River, afford good powers of limited amount.
These streams are reliable throughout the year. In cases of
extreme drought some of those heading in the coal rocks may
get low; but the others rarely, if ever, get down too low for
use.

MANUFACTURES.

There are no manufactures of any consequence in the
county. A few carding machines, to card the wool for home
use, besides the mills, are about all.

AGRICULTURE.

The principal feature in agriculture is the great grass-
producing area that covers nearly two thirds of the county.
These lands are kept perennially fertile by the decomposition
of fossil limestones, and of occasional thin ledges of felspar.

Where the lands have been in cultivation for many years, without proper rotation of crops, they show signs of exhaustion; but plaster and clover seem to restore them rapidly. Outside of the care shown by those who are trying to keep up good grazing farms, there is little effort to improve the style of farming. There are thousands of acres of land which, under a proper system of cultivation, would be capable of enriching a population of twenty times the number now living in the county. Sheep husbandry, if carefully looked after, would be highly remunerative. In the small flocks now common, and without much attention, the sheep fall an easy prey to all kinds of enemies.

Wheat, corn, oats, hay, rye, flax, buckwheat, potatoes, etc., are the chief staples, and are mostly consumed at home. There is no tobacco of any consequence raised in Russell County. Upon the great grass farms are raised much of the fine stock which supplies the English markets.

A great many fine sheep are being raised by a few individuals; horses and mules are not now the paying industries they were a few years ago.

SCENERY.

It would be useless to attempt to describe the many fine landscapes presented to the eye on every hand in Russell County.

It is made up of picture after picture of the finest views which ever came from the hand of the great artist, Nature. The beautiful grass lands, limited by the high mountains, threaded by the constant streams, which everywhere flow from bold springs, not only fill up the measure of the beautiful, but carry the conviction of the great wealth and utility of the fine lands and streams.

Fruits do well in the county. Perhaps peaches are more

certain than in the counties farther east. *Grapes* are common. *Bees* do well; and *fish*, such as black bass, the red-eye, etc., are the natural inhabitants of the Clinch and Holston rivers and their tributaries. They need protection and care, however, or they will begin to disappear.

TRADE IN CATTLE, SHEEP, WHEAT, AND CORN.

The number of cattle annually sold from Russell is about 10,500 head. Of this number perhaps 2,000 find their way to European markets, either in first or second hands. Of sheep there are about 9,500. Of wheat about 45,000 bushels, and but little corn. Of all other products, the home consumption now meets the production, except that about 450 horses and mules, in good season, find their way to eastern markets.

LINES OF TRANSPORTATION.

There are now chartered three principal lines of railway, that have some probability of being built, which will give to Russell nearly all the facilities it will require.

The Richmond and Southwestern Railway will run for thirty miles, through the Clinch River section of the county.

The Saltville and Coal Mine Railroad, if built, will run diagonally across the center of the county, from the southeast to the northwest, and would be of great value to the county, crossing the iron, marble, coal, and timber belts in succession.

The Virginia, Kentucky and Ohio Railroad has a branch road provided for in its charter, which might pursue the line of Clinch River on its way to Pound Gap. This road would afford a valuable outlet to coal and timber, toward Norfolk, without much risk of coming into competition with other coals.

TOWNS AND VILLAGES.

Lebanon, the county site, is a village near the center of the county, a short distance from Cedar Creek ; with churches of different denominations, schools, hotels, good stores, smith shops, etc. It is healthfully located. *Hansonville*, in the southwest side of the county on *Moccasin*, is a handsome little place, with a mill, stores, etc. *Honakersville*, on Lewis Creek, north side of the county, is a busy place, with a mill, stores, church, etc. *Dickinsonville* and other places in the county, such as Rosedale and Elk Garden, are convenient places of trade for the surrounding country.

PUBLIC SCHOOLS.

The public schools, according to the last report of the State Superintendent, Dr. Ruffner, are on a better footing than formerly, and will in the future be more carefully attended to.

SCOTT COUNTY.

The county of Scott differs very much in its physical appearance from the neighboring counties even of the same geological age. Its hilly character will, it is thought, finally cause it to be used principally in the culture of grapes and sheep. Along Clinch River, which flows through it from northeast to southwest, there are, however, very nice bottom lands, making many fine farms. Along the Holston River, in the southern part of the county, this condition is also observable. But Scott is important in area and in undeveloped resources. Its mineral springs are highly valuable; and its marble, ores, and coal, with its endless water power, will cause it to assume a high position under favorable conditions of transportation.

HOW BOUNDED.

The county is bounded north by the Powell's Mountain Range, separating it from Wise and Lee counties; on the east by Russell and Washington counties, and south by Sullivan and Hancock counties of the State of Tennessee.

HOW WATERED.

Scott is well watered by Clinch and Holston rivers and some of their principal tributaries, the great body of the county being traversed its greatest length by Clinch River, which has two considerable affluents in the county—the North Fork of Clinch River and Copper Creek. Holston River receives Moccasin Creek, which breaks through Clinch Mountain at Big Moccasin Gap, after traversing a large portion of the eastern part of the county.

There are, of course, many minor tributaries of both rivers. Of the Clinch River some of the most noted are Stock Creek, upon which is situated the famous natural tunnel; Cove Creek, Stony Creek, and Stanton Creek, upon which are situated Hagan's Mineral Springs.

GEOLOGICAL.

The geology of Scott is nearly similar to that of Russell County, with the exception of that part lying west of Wildcat Creek. But a close examination of the rocks near to and south of the coal area, in the northern part of the county, shows the existence of a long fragment of the dyestone ore strata and accompanying rocks; a condition which is not true of Russell County, except as to a small section known as the Big Ax Mountain.

To give a thoroughly correct idea of the geology in different parts of the county it would really be necessary to present two cross sections extending from northwest to southeast

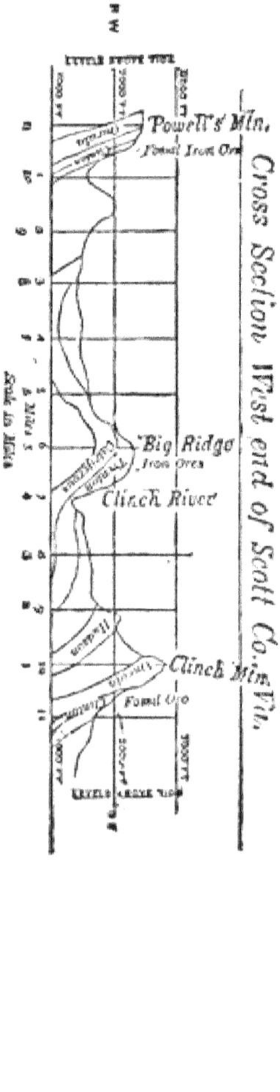

Cross Section West end of Scott Co. Va.

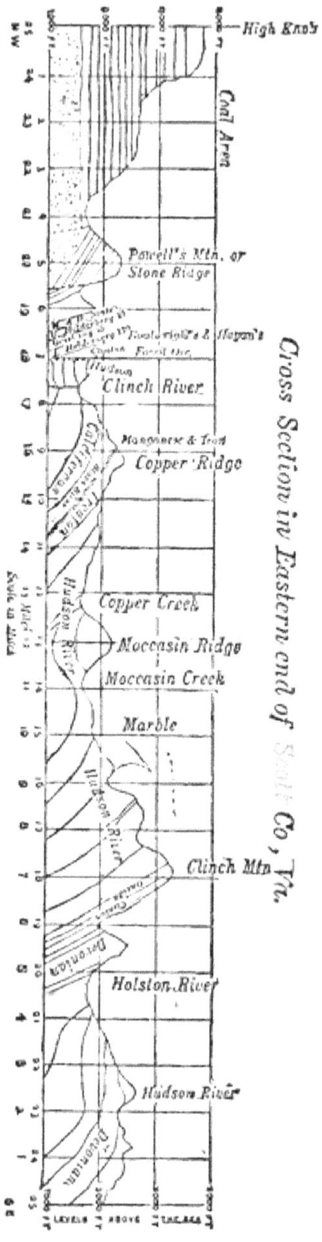

Cross Section in Eastern end of Scott Co, Va.

across the eastern and western ends of the county respectively. The eastern section could be made twenty-five miles in length, that being the width of the county at that end; while the western one would be about ten and a .half miles long, that being the best average width west of Spear's Ferry.

COAL.

The coal area of Scott County, of any value, lies in the northern extremity of the county next to the county of Wise. It is quite accessible from the Wise County or western side on the head-waters of Powell's River, as well as from the southern side, where the head-waters of Stony Creek break through the ridge forming the southern bifurcation of Powell's Mountain.

This is a much more valuable coal area than is generally believed. The veins are in better condition and lie more regularly, with much less disturbance and faults than there is apparent room for supposing. But such is the case. Seemingly this area is merely an unbroken continuation of the great coal area of Wise County under very much the same conditions. The accompanying special section will give a very good general idea of not only the regularity of the veins, but of their relationship with the neighboring strata lying south of the coal area. This coal area in Scott is about five miles broad, at its broadest, by about eighteen miles in length; and, together with the iron and manganese ores and the marble, helps to make Scott one of the most important mineral counties in the State.

IRON ORES.

Fossil or Red Ore.

The fossil ores of Scott occur in Powell's Mountain west of Flat Lick, in a low and somewhat broken ridge just south of the coal area, and in the south face of Clinch Mountain, the ores in Powell's Mountain and the ridge just mentioned being regarded as superior to the ores of Clinch Mountain. It has been often contended that Copper Ridge, in the central portion of the county, contains fossil ore also, but it has escaped notice in all the more recent examinations. Big Ridge may contain some of it, as a fragment thrown off from the main deposits, but this is extremely doubtful. It is hardly necessary to give the measures of these fossil ore veins, as they are nearly identical in all respects with the same ores in Wise and Lee counties. Thus it may safely be said that these veins aggregate sometimes a thickness of seven feet, by a length, in the county, of nearly fifty miles.

Brown Ores.

The brown ores of Scott County are almost wholly to be assigned to the Helderberg and Oriskany rocks in the south face of Powell's Mountain and Clinch Mountain; but Copper Ridge shows here and there a brown ore graduating into a red hematite, sometimes accompanied with manganese, which is no doubt in the same geological horizon with the ores of Kent's Ridge in Russell and Tazewell, and with similar ores in Rich Valley in Smyth County. Brown ores of excellent quality are also found along the length of Moccasin Ridge and Big Ridge, all of which are easily accessible, and will yield a tonnage, above water level, far beyond the power of the writer to estimate.

The brown ores graduating into a red, spoken of as being found in Copper Ridge, no doubt sometimes assume the appearance of a red ore with bright particles, leading the uninitiated to believe them to be fossil red ores. From this it may be inferred that the mistake about fossil red ores in Copper Ridge has arisen. This ore when pure is really

more valuable than the fossil ore, because it usually carries
so high a percentage of metallic iron; and it may be taken
for granted, with a degree of certainty, that in many places in
Copper Ridge, though now covered so with *débris* as not to
be discernible, there are bodies of this ore equally as great
in magnitude as those of the same horizon in Smyth, Giles,
and Tazewell counties. This horizon is about at the division
between the Trenton and Black River series, and it may be
well for the reader to remember that the same formations
follow the same ridges nearly uniformly throughout their
length in the county.

MANGANESE.

Manganese in fine crystals is found here and there
throughout the length of Copper Ridge and Big Ridge in the
county.

LEAD.

There are occasional pieces of lead ore—such as that found
near Boatwright's, on Stony Creek, four miles from Fort
Blackamore, which may lead to interesting results if properly prosecuted. Some of the pieces of galenite, picked up
where the creek, after a great freshet, had torn them from
their original position, look like the fine silver-bearing lead
ore of the West. But it would be premature to declare that
there are large quantities of the ore present in the locality
named, for much exploration and development are necessary
to prove this interesting point.

SALT.

The existence of salt may be regarded as certain in the
coal area, not more than five hundred feet below the surface;
but whether any basins will be found presenting all the conditions favorable for obtaining an unlimited supply of brine,
is not yet determined.

MARBLE.

The existence of several varieties of fine variegated marble has long been conceded to Scott County. Passing through Estillville, the county site, from southwest to northeast along the northern base of Clinch Mountain, is the line of the outcrop of large masses of both gray and purplish marble in thick ledges. The gray is tinted throughout with flesh-colored spots that render it very handsome, while the more purple is rendered remarkable from the number of large fossils with which it abounds so thickly in places, some of the remains of corals being very distinct.

The quantity of fine marble in large blocks which could be supplied from these ledges in the county, particularly near Estillville, is practically beyond the power of exhaustion, should a critical market be pleased with its coloring and style.

BARYTES.

This mineral is found occasionally in Copper Ridge and Big Ridge, but no developments of any consequence have yet been made.

FIRE CLAY.

Fire clay is found beneath some of the coal veins, but whether it will answer for the purpose of making the best kind of fire brick is not yet known.

WATER POWER.

Clinch River, before it finally passes into Tennessee on the southwest, discharges about 350 cubic feet per second; the North Fork of Holston River, about two thirds this quantity. The Clinch, having about 20 feet fall per mile throughout its length in the county, affords many admirable water powers. Holston River is not quite so favorable. Copper Creek and

Big Moccasin, and the several minor tributaries of both the
larger streams, afford water powers of any desirable grade.
It may be said that the water power of Scott County is of a
fine character, the streams being generally constant throughout the year.

TIMBER.

All that has been said of the best of the neighboring
counties with regard to timber may be said with equal truth
of Scott County.   In the coal area, the poplar, or tulip tree,
is abundant and very fine.   The chestnut oak, the bark of
which is so good for tanning purposes, is very abundant in
the county ; while, through the body of the county, black and
white walnut, hickory, maple, white oak, and other valuable
trees are abundant.

AGRICULTURE.

The agriculture of the county is not in a very advanced
condition as yet ; but that would no doubt be much improved
under the encouragement given by cheap lines of transportation, allowing the use of more profitable fertilizers.   The
cultivation of the beet for sugar making seems to be now
one of the enterprises of the county.

MINERAL SPRINGS.

The mineral springs of known efficacy and reputation are
the Holston Springs, two miles below Big Moccasin Gap, on
Holston River, and Hagan's Springs, on Stanton Creek, in the
northern part of the county.

Holston Springs, not now open to visitors, has a fine brick
hotel and a row of cabins, situated pleasantly on the north
bank of Holston River.   There are three or four springs
issuing at one place.   The largest is a spring of about
61½° Fah., slightly impregnated with iron, yielding fifteen

HOLSTON SPRINGS, SCOTT CO., VA.

(P. 284.)

gallons per minute. Near this is a limestone spring giving about eight gallons per minute, and within six feet is a sulphur spring, apparently blue sulphur. The purely chalybeate spring is not a strong one in its flow, but the water is efficacious.

Hagan's Springs are situated in a romantic spot between the mountains, having Powell's Mountain on the north. These springs are rendered attractive, not more by the excellence of the two fine springs of white sulphur and chalybeate, than the elegant house of noble dimensions and fine architecture, recently built by Mr. Hagan, the proprietor.

SCENERY.

Among the fine scenery of the State might be described several lovely views in Scott County, but one of the most attractive is the great natural tunnel or bridge in Stock Creek. Here, at an elevation of about 1,400 feet above sea level, is this remarkable natural phenomenon.

There can be no scenery more grand and imposing than that afforded by the approach to the tunnel on the lower side. An immense wall of limestone rock, of the Lower Silurian age, forms a high, beetling cliff for several hundred yards below, making a most graceful curve from south to west, then north and northeast, then curving around to south again, after passing the mouth of the tunnel, describing the curve of a gracefully turned horseshoe.

The walls of this towering cliff have been colored, in the course of time, in beautiful tints of red and sepia and brown, by the waters carrying down its face different solutions of lime, iron, and magnesia. Occasionally in some of the smaller clefts of this wonderfully beautiful cliff, cedars and overhanging pendent tufts of grass have taken hold, the cedars apparently dwarfed by the hard fight for a precarious

14

existence. The whole forms one of the most imposing pieces of scenery to be found in this part of the country. At the same time it would delight the eye, it would baffle the skill of any artist to portray it in all its features. Perhaps at no place within so small a compass can such a combination of varied and interesting scenery be found; an incomparable picture from the pencil of the greatest of all artists, nature.

Could it be opened to the tourist, health and pleasure seeker, and the weary toiler of the cities, it would be thronged every season for months, for it is really perfectly beautiful and enchanting.

A view was attempted to be taken of the lower approach from below a giant sycamore, which seemed to rear its gaunt arms as if in the vain endeavor to reach the top of the lofty cliff.

MANUFACTURES, FURNACES, AND FORGES.

With the exception of six wool-carding machines there are no manufactures worthy of the name in the county. It might be interesting to an antiquarian, who might be delving in the past history of iron making, to look up several small furnaces and forges long since out of blast, the traces of which are almost gone; but it would scarcely be a matter of interest to state them here, for they were built and operated nearly a hundred years ago, some of them; and the conditions which surrounded that remote period in our history have entirely changed. They gave way and went down under the advancing tide of Eastern supremacy in all branches of manufacture. Now the same ground is likely soon to be occupied by the successors of these manufacturers, who will be drawn to locate in a region where ores are abundant, and fuel and labor are so cheap as to make the products successful in any close competition with all the rest of the world.

NATURAL TUNNEL. SCOTT CO., VA.
Lower Approach.
(P. 210).

LINES OF TRANSPORTATION.

While both the Holston and Clinch rivers have been occasionally utilized to ship wheat, corn, etc., toward Chattanooga, the hope of the people for reliable transportation seems to be centered now in what is known as the Bristol Coal and Iron Narrow-gauge Railroad. This road has its route from Bristol on the Atlantic, Mississippi and Ohio Railroad, through Big Moccasin Gap, *via* Estillville, Spear's Ferry, the Natural Tunnel, and Flat Lick, to Big Stone Gap or Imboden City in Wise County, with a view of continuing on toward the Ohio River, through Harlan and other counties in Kentucky. The work on this road is being vigorously pushed forward, and the railroad company, with the assistance of the Tinsalia Coal and Iron Company, of Big Stone Gap, in Wise County, will soon complete the road, thus opening up to commerce the immense beds and deposits of coal and iron ore, the vast forests of timber, and the fine marble we have attempted to describe.

Fish culture, bee culture, and grape culture are alike beginning to attract the attention of enterprising men.

There can be no doubt that grape culture is suitable to Scott County in an eminent degree, and it must sooner or later command earnest attention.

ANNUAL SURPLUS OF CATTLE, SHEEP, HORSES, MULES, AND WHEAT.

Cattle.—300 stock cattle annually.
1,800 fat cattle annually.
2,000 of inferior cattle, calves, etc.
Sheep.—5,000 head.
Horses and mules.—500 head annually sold.
35,000 bushels of wheat annually shipped down the river to Chattanooga.

PRINCIPAL TOWNS, ETC.

Estillville is the county site, and the principal place of business in the county. *Nicholisville*, *Spear's Ferry*, *Fort Blackamore*, *Pattonville*, and *Osborn's Ford* are likewise places of business.

EDUCATION.

The question of education is becoming one of great importance in the county, and will assume a very healthy tone after the completion of the railroad.

LEE COUNTY.

This county has so many advantages, both mineral and agricultural, that when even so small a proportion of them as is presented here is considered it must excite surprise that no more strenuous efforts have been made heretofore to open them up to commerce. The contemplation of these and the other vast undeveloped resources of Southwestern Virginia, to say nothing of other sections of the State, kindles a feeling of deep disdain for that system of statesmanship which can content itself with the consideration of a line of policy looking to mere personal aggrandizement, through the use of all the arts of low demagoguery, while a noble State, made up in great part of such counties as Lee, is compelled to linger on from year to year in the agonies of financial dissolution; not only compromising the fair fame she has won in the past, in every department in life, but practically yielding the commonwealth, her vast resources, and a once noble heritage of an illustrious and honorable name to utter strangers. While it is not the intention of this book to go outside of the discussion of its legitimate objects, there can be no impropriety in calling the earnest attention of the citizens of Virginia to the great loss they are sustaining by failing to properly appreciate the crisis the State is now in. Without an earnest endeavor on their part to combine for the utilization of the immense resources of the State, both for the advancement of private as well as public interests, political and financial bankruptcy is certain without the intervention of foreign capital, which may be employed as a mere channel to convey away the rich fruits of Virginia's mining and general industrial resources to distant places, only partially relieving her distress at last.

Lee County can be cited as an illustration of general conditions with quite as much propriety as any other; for in its boundary lines alone are the elements of an empire's redemption from even worse conditions than those which encompass unhappy Virginia. Its fine and extensive grass and grain areas are not the least of its valuable features. In the iron ores and splendid areas of almost unparalleled coal veins, with matchless timber, there are possibilities far beyond the actual necessities of Virginia. Only one condition has been wanting for years—simply accessibility to markets; and this has seemingly been denied, as if every effort since the war had been directed to prevent rather than facilitate that most desirable end.

To the almost unlimited extent of fossil red iron ores are added valuable extensive deposits of brown ores. On a map of small scale, the coloring which shows respectively the location of ore and coal is spread over almost the same ground, and that which designates the locality of the limestone is intermingled with the iron ore and sometimes penetrates the coal areas. Nature has left nothing undone to stamp the area covered by the county as one of its most favored localities. Could it now have the number of furnaces, and mining and timbering stations to which it is so richly entitled, it would without other manufacturing establishments greatly alleviate by its tax-

paying power the burden of the State; besides making for
itself a name in iron industries which other sections might
envy, but vainly strive to emulate.

The county is over fifty miles in length by a breadth of
seventeen—air line.

Lee is separated on the north from Harlan and Josh Bell
counties of Kentucky by the main Cumberland Mountain; on
the northeast it is bounded by Wise County; on the southeast by Scott County, and immediately south by Clairborne
and Hancock counties of Tennessee. Lee is in the extreme
southwest corner of Virginia, having the State of Tennessee
on the south, and the State of Kentucky on its north and
west, and is marked at its extreme western limit by the
widely-known Cumberland Gap.

Lee is well watered by Powell's River and its tributaries.
In its southeastern and eastern corners Blackwater and Wildcat creeks flow through small sections of the county, which
are tributaries of the main Clinch River.

Powell's River, being fed by numerous limestone springs, is
a very constant stream, and toward its lower end (in the
county) is navigable through the winter months for batteaux.
It furnishes sufficient transportation to ship large quantities
of wheat and corn annually toward Chattanooga, and if a little
assistance from the general government was afforded, the
navigation of this river could be made reliable. Of this
more will be said under the head of Transportation.

It would be difficult indeed to place a proper estimate upon
the coal area of Lee County. From the examinations made

ORE KNOB COPPER MINE.

(P. 214.)

by the author, as well as those of Mr. LESLEY and others, it is
certain that Lee holds about 75 square miles of the coals
which are classed as the most reliable for quantity and most
excellent for quality in the great coal field to which it belongs.

The examinations of Mr. LESLEY and Mr. SHALER, alluded to
just now, it is true were made not so strictly in this area, but
in the exact geological continuation of the same veins and deposits on the opposite side of the Black Mountain, in a short
distance. From all the evidence, these veins lie so nearly
horizontal, without fault or displacement of a serious nature,
on both sides of Black Mountain, that the reading for one side
with but slight modifications is the reading for the other;
and this view of the case is borne out by the investigations of
careful men. Thus, Lee County, in common with Josh Bell
and Harlan counties of Kentucky, contains some of the finest
known veins of bituminous, splint, and cannel coals. Out of
the eight or nine workable seams above water level the coal
in two are known to be of that variety which will smelt ores
raw, while the quantity of the ordinary flaming bituminous
coal is without limit.

From all these examinations we are to conclude that there
are from fifteen to eighteen coal seams in all, comprised within the lower, middle, and upper coal measures. The lower
measures have the reputation of giving veins which are rarely
over three feet thick, some of which yield about one half cannel coal. This I doubt from my explorations, as, from a very
careful inspection of these lower lying beds, some of them
exceed four feet in thickness—much of the coal being capable
of smelting ores raw. The middle measures present beds
seven feet thick and over, and in these measures the thickest
cannel coal veins may be looked for.

No analysis of these coals yet inspected shows more than
four per cent. of ash, and the trials made of them recently, by

expert practical men in Pennsylvania, give them a higher
reputation in every respect, for coking and general purposes,
than any other known coals.

The North Fork of Powell's River, which derives all its
waters from this section of Lee County, flows out into the
main Powell's Valley through North Fork, or Pennington's
Gap; and it is through this gap that this fine coal area is rendered accessible throughout its extent by easy grades descending to a common point, at the above-mentioned gap.

IRON ORES.

Fossil Red Ores.

At Pennington's Forge, Pennington's Gap, these ores have
been practically proven to be of high grade. The iron made
from them has frequently been bent double while cold, without showing fracture. The localities of these ores are in
Poor Valley Ridge, Waldin's Ridge, and a section of Powell's
Mountain, which that part of the south side of the county
overlaps in taking in a portion of Blackwater Creek.

Through a great part of the county, Poor Valley Ridge,
which lies next to the Cumberland Mountain on the north
side of the county, is about eight and a quarter miles from
Waldin's Ridge, which flanks Powell's Mountain on its north
side; the latter marking for about twenty miles the southeastern boundary of the county.

These ridges, Poor Valley Ridge and Waldin's Ridge, so
well known as fossil red iron ore-bearing localities, while
maintaining the above-mentioned average distance from each
other through the middle part of the county, converge toward
each other as you approach the Wise County line, so that
when nearly at the northeastern limit of the county there
begins that great grouping of iron ores, coal measures, and

limestones characteristic of the neighborhood of Big Stone Gap.

The appended sections will show the ores in Poor Valley Ridge near Pennington's Gap; while those in Waldin's Ridge and Powell's Mountain have the same general thickness, but dip at an angle of about thirty degrees almost invariably to the southeast or south-southeast.

Again, near Boone's Path Post Office in Poor Valley Ridge, as well as in the ridges near Cumberland Gap, these ores show in measures similar to those in the section taken southwest of North Fork Gap. The ores from this place have been extensively used at Bales' or Bowling Green Forge, in connection with the brown ores mined at the forge, and have given very great satisfaction.

It would be idle to attempt to estimate the quantity of red ore in these veins. In Poor Valley Ridge the outcrop is placed at about 180 feet elevation above the water in the neighboring creeks. In Waldin's Ridge and Powell's Mountain this elevation is about 400 feet—sometimes as much as 600 feet—above water level in the creeks. It may then be safely assumed that large quantities

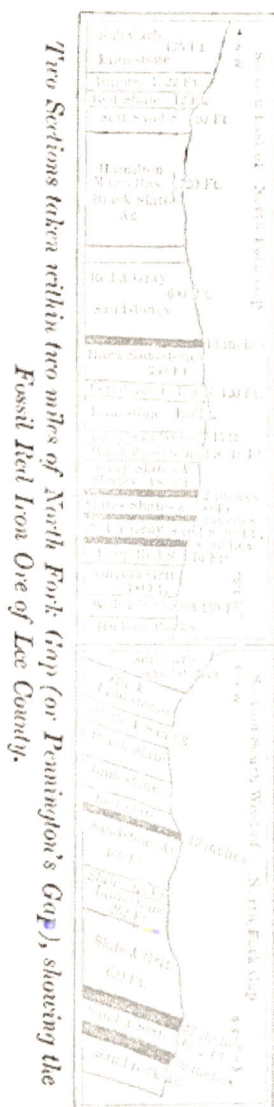

Two Sections taken within two miles of North Fork Gap (or Pennington's Gap), showing the Fossil Red Iron Ore of Lee County.

of ore will be stripped above water level. In fact, so great
will be this quantity that many years must elapse, after
regular mining operations are commenced, before it becomes
necessary to mine below water level.

Brown Iron Ore.

Brown hematite shows more conspicuously at Bales' or
Bowling Green Forge than at any other point so far as developed. Here it shows a width of twenty feet by an unknown
depth.

In the ridges to the south and east of White Shoals in
Powell's River there are considerable surface quantities of
manganiferous brown ores. The ores at Bales' or Bowling
Green Forge are in the Lower Silurian limestones, and may
be indicative of those sulphureted strata below which yield
such quantities of zinc blende farther down Powell's River in
Tennessee.

An examination of the south face of Waldin's Ridge and
Powell's Mountain reveals the presence of brown ores belonging to the Oriskany sub-epoch; but their quantity has not
been ascertained with any degree of accuracy.

LEAD AND ZINC.

The measures containing the associated strata of lead and
zinc ores do not anywhere come to the surface in Lee County.

It is probable they are as near emergence at Bales' as at
any other point. Should these strata preserve a uniform
character along Powell's Valley, the day will come when zinc
blende will be mined several hundred feet below the surface,
in the line of limestone strata passing Bales'.

LIMESTONE.

Limestone abounds in Lee County. There is very little
showing below the McLurea strata. The most of it is the

Hudson River, and in the line near the coal measures, the Sub-Carboniferous limestone. The latter is usually a fine gray, compact, and sometimes highly carbonaceous rock, well adapted to any purpose for which limestone is used, whether it be for fluxing in furnaces, for burning into lime, or for building purposes.

There are occasional bands of stone near the lower part of the Hudson River series, of such variegated colors as to have the appearance of marble. It has not been fully tested.

BARYTES.

There are occasional beds of barytes in the county.

KAOLIN,

or rather fire-clay, is found in large quantities in the coal veins. Some of it has a high character for refractory purposes. Its great quantity is an important item, should the quality prove sufficiently good for making fire-brick.

TIMBER.

The timber of Lee will form one of its most important resources in the event of transportation being supplied.

There are large quantities of walnut, maple, cedar, etc., throughout Powell's Valley. On the waters of the North Fork of Powell's River there are large boundaries of fine cherry, poplar, chestnut oak, white oak, hickory, ash, and other trees common to this latitude. The seemingly boundless forests stretch for miles unbroken. Their gigantic size, no less than their wonderful beauty and luxuriance, are calculated to impress the beholder unused to such scenes. The cedars of the lower portion of the county are wonderful for their size and number.

WATER POWER.

Powell's River and tributaries afford an immense number of fine water powers. To attempt to enumerate them would only end in showing that about every three miles of the river, and much less on such creeks as Waldin's Creek, Graybill, Indian, Blackwater, and North Fork, there are good mill sites which can employ the use of discharges varying between sixty and two hundred and fifty cubic feet per second for the river, and about twenty to fifty feet per second for the tributaries; the lesser measure for the river being up near the northeastern limit of the county.

AGRICULTURE.

The system of agriculture employed in Lee County is that which is now common throughout this section of Virginia. Many of the farms in Powell's Valley are well adapted to grass, and, consequently, are principally devoted to cattle raising. There is much of the soil in which an important constituent is a felspathic flint, perhaps a lime and soda felspar. All such soils have been found to be peculiarly susceptible to improvement by the use of plaster as a fertilizer. In growing wheat these lands have a reputation.

Lee County makes a large average yield of corn per acre throughout the county, while the wheat made is a fine firm grain, of the kind much sought after by the mills which ship their flour to southern ports.

The system of agriculture pursued, like that in use in neighboring counties, will eventually change for the better under the stimulus of increased means of cheap and abundant transportation.

SCENERY.

The scenery of Lee is picturesque, but usually softer than that of many of its neighbors. It is not wanting in such fine pictures as are presented at Pennington's Gap, Cumberland Gap, and numerous lovely stretches of woodland and mountain scenery to which the river lends an additional charm.

Some of the caves in the great limestone belt of Powell's Valley are among the most marvelous in the world for their great extent and wonderful beauty. One, a few miles from Jonesville, the county site, is said to rival the Mammoth Cave in extent, and to far exceed the Luray in gorgeous splendor of decoration.

ARCHÆOLOGY.

The archæology of the section of Lee County about Robert Ely's—between Walnut Hill and Rose Hill—is highly interesting. The Mound Builders once lived here, and have left some conspicuous marks of their existence. One of the mounds excavated a few years since by PROF. CARR, of the Peabody Museum of Boston, gave several interesting specimens of the remains, entire, of adults and children, together with ornaments of different kinds. In making these excavations, both PROF. CARR and his able assistant, Mr. Charles Johnson, of that vicinity, nearly lost their lives by the sides of the excavation falling in upon them. Over this mound these gentlemen found the remains of a walnut tree, of larger than medium size, known to have been living within the memory of citizens now residing in the neighborhood.

It is probable the walnut tree exceeded three hundred years in age. Should the remains of this tree be found over other such mounds it may be assumed that they were planted intentionally by these people. In that case, these Mound Builders would not have been extinct more than three hundred years.

MANUFACTURES.

One furnace, capable of making six tons of pig metal, was built at Cumberland Gap previous to 1861, and was rebuilt

in 1865 by a Cincinnati company, but is not now in blast.
This furnace used the fossil red ores of Poor Valley Ridge
mainly, which then yielded sixty per cent. of metal, sometimes drawing from a measure of brown ores, seemingly local,
lying between Poor Valley Ridge and Cumberland Mountain.

Bales', or Bowling Green Forge, in the western section of
the county on Martin's Creek, has been in blast for many
years, using ores from a large deposit of brown ores close
by, as well as from the fossil or dyestone ores in Poor Valley
Ridge, near Boone's Path P. O. This forge runs, generally,
only in the winter months, making about three hundred
pounds per day.

PENNINGTON'S FORGE,

Actively in operation during the winter months since 1865,
makes bar iron from the fossil red iron ores in Poor Valley
Ridge adjacent. This forge is situated on the North Fork
of Powell's River, in North Fork Gap, and has the reputation
of making a bar that will bend flat double when cold without
showing a flaw.

Outside of the furnaces and forges, the manufactories are
confined to a few carding machines and a few good tanneries,
one of which, located near Jonesville, makes so fine an article
of leather as to find a ready market among the best brands in
Baltimore, Philadelphia, and New York.

There are grist and saw-mills in the county sufficient for
present needs.

LINES OF TRANSPORTATION.

There are now no railways in Lee County. The Richmond
and Southwestern Railway proposes to extend an arm of its
great road through the county its greatest length, from Wise

County down Powell's River, *via* Jonesville, toward Cumberland Gap. This road coming, as it will, from the deep water on the seaboard, and having such fine connections with the western system of railroads, will supply remarkable facilities to the county in the development of every species of industry, manufactures, mines, and improved agriculture.

The Bristol Coal and Iron Railroad Company proposes to extend its line of road down the Powell's Valley also. This road is in course of construction, and will be pushed forward more rapidly to completion. It proposes to unite the coal and iron ores of Wise, Lee, and Scott counties with the Atlantic, Mississippi and Ohio Railroad at Bristol, and to continue its line on to the great iron deposits at Elizabethton in Carter County, Tennessee.

These lines, when completed, will in a few years place Lee County in the front rank of advanced communities.

FISH CULTURE.

Fish culture has not yet attracted the attention in Lee that is noticeable in the counties further east. The fish native to these waters are black perch, redeye, catfish, the chub, white sucker, and horny head, with some other varieties of perch, perhaps.

In a few years it will be necessary to restock the streams, as no care is being taken to keep up the supply.

Bee culture and *grape culture* are likewise neglected, there being now no great incentive to perfect either branch of industry.

ANNUAL SURPLUS OF CATTLE, SHEEP, WHEAT, ETC.

5,000 cattle, the greater proportion being stock cattle.

6,000 sheep.

50,000 bushels of wheat shipped down Powell's River in batteaux in the winter season.

The tobacco shipped annually now may scarcely be reck-
oned in the list of important items. In a few years, however,
if the spirit favoring the culture of that plant continues, the
trade in tobacco will be very large.

EDUCATION.

This subject has engaged the attention of leading private
citizens as well as the authorities, with good effect. Not only
are the public schools kept up in very good order, but there
are now good private schools, of which nothing more need
be said than that they answer all the purposes of the community, and would, under the stimulus of the prosperity resulting from increased facilities of communication and transportation, become important and of wide celebrity.

WISE COUNTY.

After exhausting a vocabulary of praise in trying to speak
of the just merits of the more strictly grass counties of this
section, it would seem that there is nothing left but to pass
over all introduction for Wise County, and begin the bare
description of its resources. Should this be done, with full
reference to those resources, there would be but little need
for any other introduction of an adulatory nature. To many
who have examined the coal veins of Wise, and others who
know something of all its resources, it would require an
exceedingly strong description to meet the full measure of
their information. No; the difficulty will be that only partial justice will be done to the county, and an apology must
here be offered for the meager account presented of its coal
and timber and other valuable features.

Wise is in the plateau of the Cumberland Mountain, having

on its northern and northwestern sides the northern bifurcation of the heavier range, and on its southern side the southern bifurcation of the same high and broad Sandstone Mountain.   It is the county which holds the widely known place in
the Cumberland Mountain called Pound Gap—a mere depression in the crest of the mountain, whose lowest point is
nearly 2,300 feet above sea level.   All railroads projected
through this part of Virginia, leading into Kentucky, generally
have Pound Gap as an objective point, thus rendering Wise
County almost always certain to be traversed either through
the middle, or near it, by the proposed routes.

HOW BOUNDED.

As just now remarked, Wise has one part of the great
Cumberland Ridge on its northern or northwestern side, and
another on its south side.   The northern one separates it
from Kentucky, and the southern, in part, from the counties
of Russell and Scott in Virginia, while a part of the county
extends over and down to Clinch River in the southeastern
section, and down to and across Powell's River in the southwestern portion.   On the east is the new county of Dickenson, and west, is the county of Lee.

HOW WATERED.

Wise County gives rise to the head-waters of Powell's River,
Gess's River, Pound River, and other forks of Russell's Fork
of Sandy River; the elevated plateau, a few miles northwardly
from Gladeville, the county site, being the divide between
the waters of Russell's Fork and those of Gess's River;
while Powell's River is separated from Gess's River by a
long, high, thin ridge, running from the High Knob in the
border of Scott, northwardly to the Cumberland Mountain.

15

The county is well watered; and those parts which partake more particularly of the limestone character—as Powell's River—enjoy never-failing streams, generally good all the year round for milling purposes; but in the strictly sand-stone districts this condition is greatly modified.

IRON ORES.

The county, extending, as it does, far enough to the south-west to take in a part of Waldin's Ridge, is able to enumerate among its iron ores large quantities of fossil red ore. Per-haps the quantity of fossil ore is more than equal to that of all other iron ores known to be accessible in the county. While it is true that there are considerable quantities of brown iron ore, of that variety which is supposed to result from the decomposition of iron carbonate, or black band ore, there have not been sufficient developments as yet to set-tle either the question of the actual derivation of this brown ore or its approximate quantity. Near the tops of the ridges, throughout almost the whole extent of the county, handsome fragments of brown ore may be picked up; but to positively assert that they are derived from the decomposi-tion of carbonate more than from sulphuret, would be taking too much for granted; and, unfortunately, there have been no well-directed efforts at showing either the number, thickness, or true character of the veins from which these surface ores are derived. That they will, finally, be found of the gen-eral character of the iron ores of the same coal measures at other known points can scarcely be doubted. But it is in the neighborhood of Big Stone Gap, toward the southwestern corner of the county, where the most favorable conditions exist for the development of iron industries. At that point, a large proportion of such brown ores as that section of the coal field will yield may be brought down to a common point,

Cross Section at Big Stone Gap,
Wise County, Va.

Plateau of The Cumberland Mountain

Fossil Ore Measures Enlarged

NOTE.—The enlarged Section of Fossil Ore is from the ridge north of the fault at north base of Watkins Ridge.

toward which both the fossil red iron ore of Waldin's Ridge
and the coal will all naturally gravitate, their position in
the ground being within a few miles of each other.

An examination of the map accompanying this book will
readily show these conditions more clearly than any description can; and it is much to be regretted that the brown iron
ores of that end of Powell's Valley—not in the coal measures
—should have been so little developed so far as to leave the
question of their quantity still in doubt. Should this question once be settled satisfactorily, it may readily be inferred
that the large quantity of fossil ore would find a most convenient and effectual counter-check for its possible impurities in an abundance of pure limonite. In fact, it is grati-
fying to know that a company of experienced Pennsylvania
iron men are now engaged in settling this question in a most
practicable manner.

The accompanying section is submitted more to give an
idea of relative positions than as a thoroughly accurate reading of the geology of that immediate locality. In it may
also be gathered some idea of the truly fine exposures of the
coal.*

COAL.

As may be seen from some of the measures recorded below in the foot-notes, Wise County is justly celebrated for its

* In the immediate vicinity of Big Stone Gap, on Pigeon Creek and
Looney's Creek, mentioned in the cross section, the field notes read as follows :

At 2,075 above sea level, barometer 83 Fahrenheit, on a branch flowing south
into Pigeon Creek, at a point 1½ miles from Big Stone Gap, on lands claimed
by MATHIAS KELLY, now owned and soon to be operated by the Tinsalia Company, found coal about 11 feet thick on a floor of slate, roof of slate 4 feet,
overlaid with sandstone. The vein beginning at the floor measures nearly 4
feet of bituminous coal, then 4 inches of slate, then nearly 4 feet of coal
(bituminous), then 4 inches of slate, then nearly 4 feet of coal, mainly splint,
up to the roof.

coal veins. Not only on the headwaters of Powell's River, but on Gess's River and tributaries, Pound River, and on the streams flowing into Russell's Fork of Sandy River, a series of rich veins of bituminous, splint, and cannel coals are found lying in nearly horizontal beds, and outcropping in such a manner, in many places, as to plainly show their character and thickness, with but little labor on the part of the prospector. Gess's River being regarded usually as the line which most projected through railways are likely to take, it may be of interest to look closely at the facts elicited from the exam-inations which have been made up and down that stream. It is commonly supposed that the 4 to 6 feet vein, which shows about water level in the river near Gess's Station, is the lowest of that series. It may not be improper to suggest that this is really the second vein of consequence, number-ing from below, in the lower measures, the first vein—of about 11 feet thickness—being under it less than 80 feet, while above this No. 2 vein of 4 to 6 feet are three other principal veins, respectively (as you ascend) 4 feet and 5 feet of bituminous

---

At the Looney's Creek opening, three miles north from Big Stone Gap, barometer 1,940 feet, weather damp, thermometer 74° Fahrenheit, found 68 inches of coal with a parting of 5 inches of slate near the top. Of this coal 10 inches was splint coal, floor of slate, roof of slate, 8½ feet thick up to sand-stone.

At other points further toward the head of Powell's River, readings of the coal veins were obtained. On Black Creek, a tributary of the right-hand Fork of Powell's River, seven miles west-southwest from Gladeville, saw a coal vein 6 feet thick, floor of clay and slate. On Rocky Fork of Roaring Fork, 2,040 feet, barometer 85° Fahrenheit, found a coal vein 7 feet 2 inches thick, having thin seams of splint coal, with two partings of slate 4 inches each; roof of black slate 15 feet thick overlaid with sandstone; floor of slate full of fossils, leading down 20 feet to another vein of coal 18 inches, which is bedded on sandstone. This is 8½ miles west of Gladeville. Also here found a deposit of red iron ore of unknown extent, of peculiar structure, honeycombed, red, light, but some of it a close purplish gray, and heavy, very likely a decomposition of carbonate of iron.

coal, with the upper vein 4 feet of pure cannel, about 750 feet above the level of No. 2. There is scarcely any need to go into a more particular description of these veins. They extend with but little intermission over nearly all the county, except that they are eroded by the streams, and the cannel coal, being nearer the crests of the higher ridges, does not occupy more than half the area that the other veins do.

The reading of these measures is fully sustained, not only by the careful examinations made by the author, but MARTIN COYELL, Esq., an experienced and reliable mining engineer, gives very much the same readings, to say nothing of the results of earlier examinations made by MR. J. P. LESLEY and others.

The industrial value of the coals of Wise County can scarcely be overestimated. The day must surely come when they will be so largely employed in industries carried on in Virginia that their great quantity, purity, and excellence must make them perform a great part of that giant task of relieving Virginia from her pecuniary embarrassments. Could a large proportion of the Virginia capital, which is now being sent outside of the State annually for fuel, be invested in Wise and the neighboring coal counties, it would be but a short time that, from this source alone, the tax-paying power of those communities would be so increased as to materially alleviate the financial distress of the State, if not go very far toward its entire extinguishment.

It may be as well to say, before dismissing this important subject, that all analyses which have been made of the coals of Wise County show a low percentage of ash, and a high percentage of combustible matter. And it would be well also to say, that the coals along the Pound River and the tributaries of Russell's Fork of Sandy River, are of the same general character, thickness, etc., as those already described.

We quote the following extract from the report of PROF.

JNO. J. STEVENSON, professor of geology in the University of the City of New York:

"The coal area is well opened up by the several branches of Pigeon Fork as well as by Looney Creek, Calahan's Creek, Canepatch Fork, and other tributaries to Roaring Fork. At the immediate base of Stone Mountain the dip is very abrupt, but it quickly diminishes, so that on the opposite side of Roaring and Pigeon Forks it is but two degrees. The decrease is rapid, and within a mile the inclination of the beds is little more than sufficient for convenient drainage.

"Though the streams have cut deep channel-ways so as to bring the coals above water level in a great part of the area, yet the general section was obtained with some difficulty, as there has been little development and the whole surface is covered by dense undergrowth. The intervals in the section may require some modification as they were measured with an aneroid barometer.

"The succession as obtained is as follows:

1. Not examined in detail...............300'
2. Coal bed............................ 0' 4"
3. Interval............................ 30'
4. Splint coal bed..................... 1' 5"
5. Interval............................'115'
6. Coal bed............................ 0' 6"
7. Interval............................ 70'
8. Coal bed............................ 2'
9. Interval............................ 14'
10. Splint coal bed.................... 1'
11. Interval........................... 90'
12. Coal bed........................... 0' 4"
13. Interval........................... 30'
14. Coal bed........................... 0' 10"
15. Interval........................... 65'

16. *Coal bed*........................... 7 3″
17. Interval........................... 35′
18. *Splint coal bed*..................... 3 6′
19. Interval........................... 60′
20. *Splint coal bed*..................... 1 6″
21. Interval...........................120′
22. *Coal bed* ....Blossom.
23. Interval .... 70′
24. *Coal bed*........................... 1′ 2″ to   2″
25. Interval........................... 28   to 20′
26. *Coal bed*........................... 15   to   2′
27. Interval........................... 50′   to 45′
28. *Coal bed*........................... 8′ 5″ to 6′ 9″
29. Interval...........................206′
30. *Coal bed*........................... 2′
31. Interval........................... 50′
32. *Coal bed*........................... 2′
33. Interval...........................100 ?
34. *Coal bed*........................... 5′
35. Interval to conglomerate............ 80′ ?

      Total...........................1,518′

" The exposures are most satisfactory on the tributaries of Calahan's Creek, a tributary to Roaring Fork. The higher coals are of no importance.

" *Coal bed*, No. 16, is well shown on Preacher's Run at about two miles above where it enters Calahan's Creek, where it shows

Cannel................................. 2′
Alternations of *coal* and shale................... 1 10″
Coal................................. 3 5′

" The *cannel* is compact and of by no means inferior quality, though it may contain as much as 12 per cent. of ash. The *bituminous coal* at the base of the bed is good and seems to be quite free from sulphur. The same bed was seen near the head of Calahan's Creek, where it is worthless, being in three benches, respectively 5, 14, and 5 inches thick, separated by 10 and 7 feet of shale.

" *Coal bed*, No. 18, was seen on Preacher's Run and near the head of Calahan's Creek. At the former locality, near two miles above the mouth of the Run, it is 3′ 6″ thick, and is *splint coal* of superior excellence. The roof is a hard sandstone. On Calahan's Creek it is but 18 inches thick and is worthless. Fragments of the coal from this bed were seen on Looney Creek, but the bed itself is concealed.

"No. 20 was seen on Preacher's Run and Looney Creek, but its place is concealed at all other localities examined. On Looney Creek it is 4′ 6″ thick, and is well exposed in the bank of the stream at about three miles above its mouth. The coal is clean and evidently an excellent splint. The same bed is shown at the forks of Preacher's Run, two miles above its mouth, where it is from 3 feet to 3 feet 6 inches thick and is a good *splint coal*, though the quality seems to be hardly equal to that of the coal on Looney Creek. The blossom of this bed was seen near the crest of the hill overlooking the mouth of Looney Creek. At all localities the bed has an excellent roof and it can be worked very easily.

"*Coal bed*, No. 26, is persistent but exceedingly variable. It has been opened on Pigeon Fork at probably two thirds of a mile above the mouth of the stream, where the structure is

Coal	6′ 10″
Shale	0′ 10″
Coal	2′ 5″
Shale	0′ 2″ to 5″
Coal	4′ 6″
Total	14′ 10″

A similar structure is shown on the first branch of Pigeon Fork. At both exposures the upper division is slaty and inferior; the middle division is prismatic and cokes well; the lower division is a good splint coal, though toward the base it becomes somewhat slaty. On Looney Creek, only the middle and lower divisions are present, the upper one having been removed by a *horseback*; the coal of those divisions shows the same features as on Pigeon Creek, though the middle d vision is better, and is packed in sacks to a distance of several miles for the use of blacksmiths. On Kelly's Run and Church House Run, branches of Calahan's Creek, the bed is thin, probably not more than two feet, while on Preacher's Run it is but two inches. It is 10 inches thick on Calahan and the coal is poor. An imperfect exposure was found on a branch of Roaring Fork, at, say, a mile above the mouth of Calahan, where the bed seems to be very thick.

" *Coal bed*, No. 28, is the most important and least variable of the whole series. It was seen on Pigeon Fork and one of it branches, on Church House, Kelly's and Preacher's Runs, as well as on Lewis's Branch of Roaring Fork at about three miles above the head of Big Stone Gap. No exposure was found on Looney Creek or on Calahan, the bed being concealed at the level of those creeks. On Church House Run an opening shows

Coal	3	10″
Shale	0	5″
Coal	3	6″
	7	9″

The top of the bed for 11 inches is a hard slaty splint, but the remainder of the upper bench is very soft and some of it has a prismatic structure. The lower bench is less soft. An exposure near the mouth of Preacher's Run shows:

At a mile further up the creek the bed shows the same structure, but is only 6 feet 9 inches thick. A small rick of coke made here proves that the coal will burn into a compact silvery coke of great strength. On Kelly's Fork the bed is somewhat larger, the benches being 4 feet 5 inches and 3 feet 9 inches respectively, separated by three inches of shale. The bed is nearly 8 feet thick on Lewis's Branch of Roaring Fork.

"The lower coals of the series are unimportant. The lowest coal was seen only in the bed of Pigeon Fork very near the base of the Stone Mountain, where the dip is nearly 60 degrees.

"The intervals between the coal beds are occupied almost wholly by sandstone, most of which is compact. Limestone is almost wholly wanting.

"THE IRON ORES.

"Three beds of the *fossiliferous* ore were found in the Poor Valley, between Poor Valley Ridge and Stone Mountain. The highest one is exposed in the bank of Powell's River at about 200 yards above the ford leading to the Big Stone Gap, where it is from 5 to 8 inches thick. This bed was not traced, but it clearly follows the bottom for a long distance, and lies east from the river below the ford leading to Cedar Gap Church. The dip is north-northwest at 55 degrees.

"The second bed was seen on Mr. Horton's property at

about a mile and a half from the Big Stone Gap. There it shows

Hard ore...................................3
Medium ore...............................0' 8"
Hard ore..................................1
Soft ore..................................2 6'
            ————
            7' 2"

The strike here is not far from north 40° east Mag., and the dip is 45 degrees toward the northwest. The exposure is continuous for a long distance, and the soft ore at the bottom is very good. Doubtless the whole of the ore from this bed, excepting the very hard ore from the top, could be utilized in a furnace.

" A third bed was seen near the Cedar Gap Church, but the exposure is somewhat indefinite, and the soft ore seems to be present in comparatively small quantity.

" The lower two beds are present on the north side of Waldin's Ridge, but the ore was not found in place. Large fragments of it occur plentifully on the side of the mountain, and the beds could be discovered without serious difficulty.

" The *fossiliferous* ore is present on the south of Waldin's Ridge. A bed was found on the Preston tract at less than four miles from Big Stone Gap, which is 25 inches thick and dips eastward at but 10 degrees. Its outcrop can be followed round and on both sides of a low ridge separating two hollows, and the ore can be mined with equal ease on the side of a similar ridge lying immediately west. It is not too much to assume that there are 500 acres of good ore land on this tract. The most of this lies under thin cover, and much of the ore can be obtained by stripping. Drifting, however, can be performed easily, as the ore can be opened so that the mines will drain themselves.

"At a little way further southwestward along the face of the ridge another and possibly a lower bed of the ore was seen, which shows

Soft ore . . . . 2'
Hard ore . . . . .2' 2''
$$\overline{}$$
4' 2''

overlying one foot of ferruginous shale. This rests on a flaggy sandstone, and the roof of the ore is a clay shale which can be removed easily. The soft ore is double, the upper part for somewhat more than one foot being extremely soft, so that it could be removed with a spade, while the lower part is made up of the ordinary ore. The hard ore is very hard and shows few fossils. This bed is hardly so favorably situated as the other, for the dip is directly away from the exposure. Still a drift could be so arranged that an enormous quantity of the ore could be won without resort to artificial drainage. The hollow is a long one, and the ore can be stripped to a distance of fifteen feet from the outcrop without much trouble.

"The fossil ores occur also on the south face of Powell's Mountain south from the Ward's Mill or Slemp's Gap. This area is known as the Kane survey, and, by the proposed railroad, is about fourteen miles from the Big Stone Gap. Three beds were seen there. The top one is thin, very silicious, and of little value. The second bed has been opened so as to show its structure, which is

Ore . . . . 2'
Shale . . . . 1'
Limestone and ore . . . . 0' 5''
Shale . . . . .0' 4''
Ore . . . . 0' 5''
Shale . . . . .0' 4''
Ore . . . . .1'

The ore is good and would doubtless repay the cost of drifting; but it is leaner than the ores of Waldin's Ridge and
Poor Valley, which have been described already. The third
bed is at about 50 feet below this and shows

Very soft ore...................................1´ 3´´
Hard ore......................................1´ 6´´

The ore is evidently lean, but it can be mined very cheaply as
the face of Powell's Mountain is cut by many deep and long
ravines, which expose the beds.

"The Lower Helderberg rocks along the valley between
Waldin's Ridge and Powell's Mountain, as well as the valley
between Stone Mountain and Powell's Mountain, yield some
brown hematite. Two horizons of this ore were seen, but no
estimate respecting the quantity can be made until after sys-
tematic exploration of the property. The ore is very good
at some localities, and much of it was mined on the Collier
tract, now belonging to the company, which was reduced in a
furnace below Ward's Mill.

" LIMESTONE.

"The lower Carboniferous limestone is shown in sharply
dipping cliffs on both sides of Big Stone Gap, and extends
along the south face of Stone Mountain, from the Little Stone
Gap westward to beyond Pennington's Gap, near the Tennessee line. It is shown as a cliff on the north side of Powell's
Mountain from the Little Stone Gap to the Ward's Mill Gap,
where the North Fork of Clinch River breaks through the
mountain.

"The Helderberg limestones, which are well shown in the
Poor Valley at but a little way above the Cedar Gap Church,
and in a line of low hills almost directly opposite the Big
Stone Gap, are for the most part silicious; but near the top

of the lower division of the limestone a thin bed was seen which is of good quality.

"The Trenton limestone is not reached within the area controlled by the company, but is well exposed for miles along the north side of Waldin's Ridge west from the Cedar Gap.

"GENERAL SUMMARY OF THE RESOURCES.

"The available coal seams are four—the two *splint beds*, the 15 foot and the 8 foot bed. Eventually the *cannel* bed may become valuable, but at present it may be omitted.

"The splint coals lie well up in the series, and the extent of area which they underlie cannot be determined by a hasty examination; but it can hardly be less than 5,000 acres, within the area drained by Calahan's Creek, and lying between that stream and Looney Creek. These beds yield a coal of very superior quality, as appears from the following analysis by Mr. A. S. McCreath, chemist to the Geological Survey of Pennsylvania. The sample was taken from the upper bed.

Water... 0.880
Volatile matter 37.580
Fixed carbon 58.059
Sulphur 0.406
Ash.. 3.075

"This coal is sufficiently pure to be used raw in the furnace, the percentage of ash and sulphur being unusually low.

"These two beds will yield not far from 8,000 tons per acre. Estimating the area of easily mined coal at 5,000 acres, the yield will not be far from 40,000,000 tons.

"A specimen was taken from the middle division of the 14 foot bed, which yielded the following upon analysis by Mr. McCreath :

Water 1.610
Volatile matter 38.850
Fixed carbon 57.879
Sulphur 0.771
Ash 0.890

"As a gas coal this can hardly be excelled. It cokes readily, and the coke, if compact enough, should be of excellent quality. No analysis of the lower division was made; it is a splint coal which will be well worth working. This bed is uncertain in its occurrence, and seems to attain its chief importance west of the line of Preacher's Run, having been found of large size only on Looney Creek and on the waters of Pigeon Fork.

"The 8 foot seam being the most persistent and least variable of the whole series, and being likely to prove the most valuable, as its coal yields an admirable coke, I took samples from all the benches and directed that they be analyzed as one. The result, as obtained by MR. McCREATH, is as follows:

Water 1.380
Volatile matter 35.920
Fixed carbon 60.591
Sulphur 0.594
Ash 1.515

"In reference to these coals MR. McCREATH says:

"'The above analyses speak for themselves, and indicate coals of remarkable purity.'

"This 8 foot seam will yield a coke with *considerably less than 3 per cent. of ash, and with but little more than five tenths of a per cent. of sulphur.* Such would be a marvelously rich coke, the percentage of fixed carbon being *somewhat more than* 96. The Connellsville coke has somewhat less than 90 per cent.

of fixed carbon, the ash is between 9 and 10 per cent., while the average of sulphur is about eight tenths of a per cent. The coke from this bed is better than that from the Oxmoor Works, in the Cahawba basin of Alabama, which shows

Fixed carbon 93.253
Sulphur 0.601
Ash 5.380

"A small rick of coke was made from the 8 foot seam on Preacher's Run. Though carelessly made in disobedience of the instructions, the coke proved to be of great strength. 8,000 acres are underlain with this bed between Looney Creek and Roaring Creek, which will yield 64,000,000 of tons without resort to artificial drainage.

"The *iron ores* of widest distribution are the *fossil* ores. No effort had been made previous to the time of my examination to secure a good exposure of these ores, and the beds were traced by their outcrop. Some difficulty was encountered in obtaining specimens which had not been exposed for a long period to the weather, and collections were made only on the south side of Waldin's Ridge. These were sent to MR. McCREATH, who gives the following as the results of the analysis:

	No. 1.	No. 2.
Metallic iron	27.960	52.600
Sulphur	0.024	0.018
Phosphorus	0.064	0.116
Insoluble residue	55.015	18.140

"No. 2 is from the company's property on the Preston tract, and is precisely like the ore which the company has secured on the Horton tract in the Poor Valley. It is an excellent ore, and is present in enormous quantity. It will give an iron with but .22 per cent. of phosphorus. The analysis,

16

No. 1, is from a tract not far from Big Stone Gap. The sample is not a fair specimen of the ore, as it contains many rounded quartz pebbles; but no other specimen could be obtained from the outcrop.

"A sample of the brown hematite, found also near Big Stone Gap, Mr. McCreath finds to be as follows:

Metallic iron................................52.550
Sulphur.................................... 0.037
Phosphorus................................ 0.051
Insoluble residue.......................... 7.840

"This is an excellent ore, and is well adapted to the manufacture of Bessemer pig.

"No analyses were made of the limestone at the mouth of Big Stone Gap. It is an excellent rock, burning into fine white lime, and seems to be almost free from silica.

"All these materials are in close proximity to each other, and there are few localities more favorably situated for the manufacture of iron. Coke can be brought from the pits, barely two miles away, on a down grade; the *fossil* ore is but little more than one mile away, while some of the ore is distant but three miles, and is directly on the line of the railroad now in course of construction. The *splint coal* can be mined cheaply, and it can be used with the coke in the furnace. The purity of the coke and of the splint coal would aid in making an iron of exceptional excellence. The cost of making iron here may be estimated as follows:

2.25 tons of ore at $1.40 per ton.............$3.15
2 tons of coke............................. 2.00
1.5 tons of limestone...................... 0.60
Labor at furnace 1.50
Repairs and incidentals.................... 1.00

"Making a total cost of $8.25 per ton, which may prove

somewhat in excess of the absolute cost of material and labor.

"A charcoal furnace could be managed to good advantage in connection with the larger coke furnace. The *fossil* ores have long been used in the manufacture of cold-blast charcoal iron, and under that treatment in the Poor Valley they produce a pig with but .17 per cent. of phosphorus. An enormous quantity of excellent coaling timber is found on these lands, and the charcoal could be obtained at not more than three dollars per hundred bushels. A small furnace of this kind, though not affording much direct profit, would be of advantage indirectly, since in the manufacture of the charcoal a large part of the land could be cleared and fitted for occupation by farmers. The region is admirably adapted to stock raising, while much of it lies in such a position that vineyards could not fail to be profitable."

LEAD.

Lead is sometimes found in small quantities in a stratum of limestone belonging in the coal measures, but not in sufficient quantities to attract the attention of miners.

SILVER.

Silver is now and then observed in the quartz pebbles composing the conglomerates, and may be found at some points in respectable quantities. It is hardly to be expected that any reliable data will ever be gathered upon which extensive mining operations will be based, looking to silver mining as a profitable source of investment.

BUILDING STONES.

The sandstone ledges in a great many parts of Wise resemble those of Buchanan County in the cheapness with

which they can be quarried and made ready for use in any
desired shape or size. Much of this stone is so soft in the
quarry as to readily admit of being cut with a knife, but
upon sufficient exposure, becoming quite hard and durable.

TIMBER.

Upon this subject too much cannot be said of the variety
and beauty of fine woods in the county.

Cherry is very common in the Big Black Mountain, toward
the northeastern side of the county. Its abundance is one
of its great features. Of all other trees known to the latitude—prominent among which is poplar, or the great American tulip tree—there is such great abundance as to appear
totally beyond the possibility of extermination. The girth
and length of trunk of the trees in these vast and almost
boundless forests are surprising. Oftentimes the poplars
are found 6 and 8 feet in diameter and sometimes larger,
with long straight trunks 75 or 80 feet without a limb. The
cherry trees are also surprisingly large and beautiful.

AGRICULTURE.

Agriculture in connection with sheep raising would yield a
more certain return on a large scale than any other branch
of farming. That section of the county on Powell's River
below Big Stone Gap, mostly limestone, is well adapted to
any description of farming, being similar in this respect to
that part of the county in Clinch River near Wheeler's
Ford; but the greater part is composed of sandstone ridges
and plateaus, which seem to sustain sheep better than cattle;
besides from their steepness in many places presenting poorer
facilities for all kinds of farming than the more level lands
on the river do. Notwithstanding this fact, these lands pro-

duce corn well, appearing to be very fertile. Sweet and Irish potatoes do well.

It may well be inferred, that without railway or other improved modes of transportation, there has been but little encouragement to engage in manufacturing enterprises. At Big Stone Gap a company has recently secured large areas of coal, timber, and iron ore lands, with a view of erecting furnaces for the production of metal on a large scale ; but it is not supposed that the works will be put actively into operation until the Bristol Coal and Iron Railroad, and the Richmond and Southern Railway are built.

At Three Forks of Powell's River, just below Big Stone Gap, there is one of the very few pottery establishments in Virginia. It is true it is not on a large scale, but if kept up until the contemplated railway lines are built, there is no reason why it could not largely increase its working capacity. MR. WOLF, the proprietor, deserves much credit for the excellent drain pipes his factory makes, besides pottery of various patterns. It is very probable that the carboniferous strata close by furnish the clay used. Drain pipes are sold there at 15 cents per foot.

The streams, except Powell's River, being rather inconstant during much of the year, there are few other manufacturing enterprises worthy of note, except grist and saw-mills.

SCHOOLS.

Wise County depends mainly upon her public schools for instruction, and the population as yet being sparse, the success which attends the system in other parts of the State is not so marked in Wise. This, however, is a small matter ; for as soon as the mining facilities of the county are once

246

utilized to any extent, the consequent increase in population
and revenues will enable the Superintendent of Education to
place the county upon an admirable footing. The same may
be said of the other counties contiguous.

TRADE IN CATTLE, ETC.

Wise County sells annually about 2,000 head of stock cattle
and 3,300 head of sheep.

There is still a considerable trade in ginseng, herbs of different kinds, and wool. It will not be forgotten by the reader
that there are no developments yet of the coal and iron of
this county, hence the trade of every kind is very small. The
culture of the grape and of bees is a question of much interest in certain parts of the county.

DICKENSON COUNTY.

Dickenson is a new county lying between the counties of
Buchanan and Wise, and was formed by Act of Legislature,
session 1879-80, from the two counties named.

It is a small county, as may be seen by referring to the
map; but it has a wealth of fine bituminous, splint, and cannel coals, unsurpassed by the same area anywhere. The
timber is truly magnificent, and stretches in an almost unbroken forest all over the county. The poplar trees are fine
and very numerous. There is also a great deal of walnut,
white oak, and other valuable woods. The county produces
corn and sweet potatoes well, and ranges a good many cattle
and sheep. The scenery of the county is very imposing here
and there, especially that in the deep cañon about the breaks
of the Cumberland Mountain, in the lower or northern end of
the county, on Russell's Fork of Sandy River, by which stream
the county is chiefly watered.

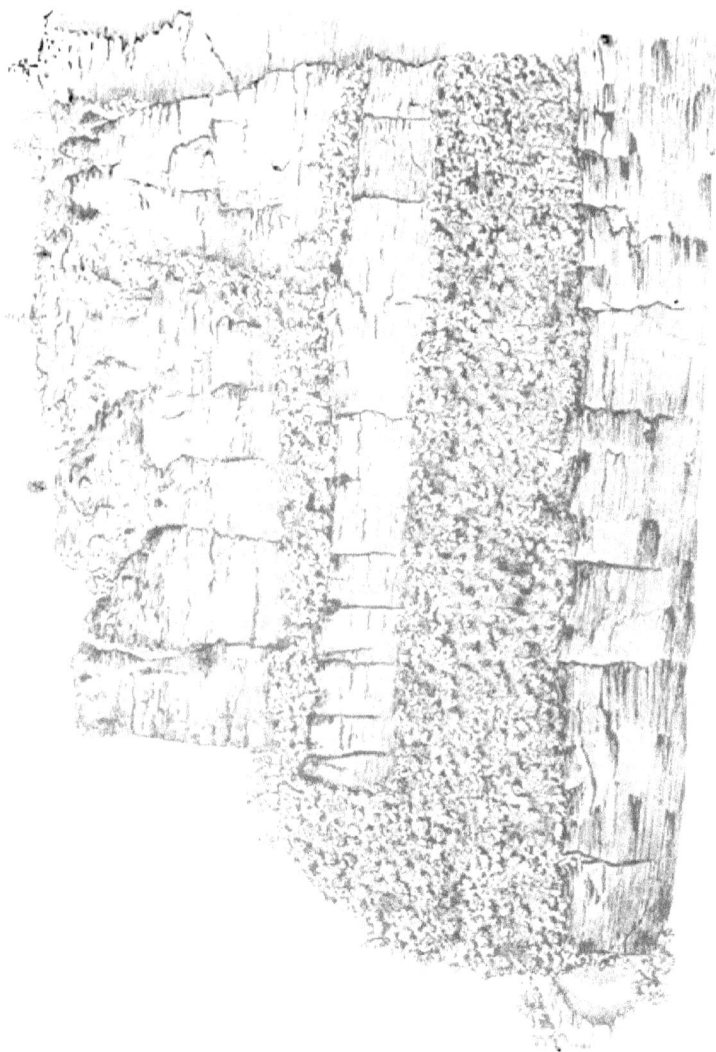

THE TOWERS, RUSSEL'S FORK OF SANDY RIVER.

(P. 246.)

BUCHANAN COUNTY.

The description of this county will be mostly confined to an account of its coal and timber. Its land is almost without exception sandy. In many parts the loamy character of the soil renders it very fertile ; but greater care has to be taken to keep it up to its original strength than the limestone soils. The salt of Buchanan would prove an important item if developed. The rocks appear to dip in such a way as to form several basins, fully capable of holding sufficient of the drainage from the salt-bearing series of rocks for all purposes of salt-making on a large scale. The scenery of this county here and there is grand and beautiful in the extreme. At the breaks of the Cumberland Mountain on Russell's Fork of Sandy River, this is the case in an eminent degree, presenting a piece of scenery rarely met with this side of the Rocky Mountains.

HOW BOUNDED.

Buchanan is one of the extreme border counties of the State, in the great plateau of the Cumberland Mountains, having Kentucky on the north and northwest, West Virginia on the east, Tazewell and Russell counties, Va., on the south, and Wise County, Va., on the west. The main Cumberland Ridge is its northern and northwestern boundary line ; its southern the Sandy Ridge or southern bifurcation of the Cumberland Mountain ; its eastern line is on a great dividing ridge between the waters of the Louisa and Dry Forks of Sandy River, and its western line is an irregular one, soon to be disturbed by the laying off of the new county of Dickenson.

HOW WATERED.

The southern boundary line of the county, in following the general crest of the Sandy Ridge, divides the waters of Sandy

River from those of Clinch River, Sandy River having two of
its branches—the Louisa Fork and Russell's Fork—to take
their source in the southern part of the county, and flow
northwardly into Kentucky; the former watering the eastern
part of the county, and the latter the western portion. These
two streams are constant the year round, but are too low in
summer, except at their lower ends, to be relied upon to run
machinery. There are some grist-mills on them and their
tributaries, however, which seem to run with some regularity, except in very dry weather.

GEOLOGY.

As to the geology of Buchanan, the rocks, as they now
show on the surface, indicate the Sub-Carboniferous as the
true period; yet it is singular, that at the bottom of the
series here there should be a vein of coal 10 feet thick.
Not more singular than that Russell County, near by, should
show one 9 feet 4 inches thick at the base of the series
there. There are so few disturbances or faults worthy of
notice in the rock formation in this county, that a description
of the series of rocks, veins, etc., at one point may be taken
as a fair reading of nearly the whole. There are variations
in the strata, it is true, but for Sub-Carboniferous measures
they run with surprising regularity for miles. The fossil
remains of plants and trees are quite common here and
there, such as ferns, rushes, grasses, lepidodendra, sigil-
lariæ, etc. The dip of the rocks, except in the extreme
northwestern part of the county, is but gently inclined from
the horizontal. The rocks are almost wholly sandstone,
shales, and slates, interstratified with veins of coal and occasional thin bands of iron ore, either as sulphuret or carbonate. Now and then you meet with limestone, as on the
divide between the head-waters of Big Prater Creek and the

waters of Russell's Fork, and at Countz's on Lick Creek. This limestone sometimes shows lead ore, but it is doubtful whether that will ever be found in sufficient quantities to justify working.

IRON ORES.

The iron ores of Buchanan County, as they appear on the surface, are mainly *brown ores*, very often alluded to as hydrated peroxides. They result from the decomposition of carbonates and sulphurets, and may be found near the crests of nearly all the ridges in the county. Toward the western side of the county there seems to be the greatest quantity.

It does not appear from the amount showing on the surface that the undecomposed veins are very thick; perhaps in some instances eighteen inches or two feet; and so few are the developments that it is now impossible to tell their true character. There are no other ores of iron observable in the county, except here and there a piece of undecomposed sulphuret.

No manganese worthy of notice has been shown by actual development; but this is not saying there are no manganese ores in the county.

COAL.

Besides timber, coal is the principal feature in the county. The heavier workable measures seem to lie near the tops of the hills, except on Connoway Creek, near the northern line of the county, where the accompanying section was taken. There the same veins, that show at somewhat higher levels farther south, are nearer the creek. The coal is almost without exception flaming bituminous in all the veins. Toward the western boundary of the county it assumes the appearance of that variety which is used for smelting iron ores raw. This series of veins lying nearly flat throughout the

Vertical Section-Conway Creek		
Strata	Kinds & Thickness of Rock & Coal	
2000	Top of Ridge Sandstone	
1950 / 1900	Slate — Coal 9 in.	
1800	Slates & Sandstones	
1750 / 1700 / 1600	Sandstone — Coal 4 Ft. — Slate (Blue)	A
1500	Slates & Sandstones	
1400 / 1375 / 1300	Slate — Coal 31 in. — Slate — Slate & Sandstones	B
1200	Coal 4 Ft. 3 in.	C
1175 / 1100	Coal 5 in. — Coal 3 Ft. 10 in. — Slate & Sandstone	D
1000		
900 / 860 / 800	Slate — Coal 22½ in. — Slate — Sandstone	E — Water Level — Loutsa River
730 / 700 / 600	Coal 10 Ft. — Shales & Sandstone — Conglomerate — Thin Coal — Sandstone	F
500	Conglomerate	
400	Sandstones & Shales — Mountain Limestone	
300 / 200	Shales & Slates / Sandstone — From the mountain limestone 1100 feet approximately down to the Coal Oil Rock	

Buchanan County.

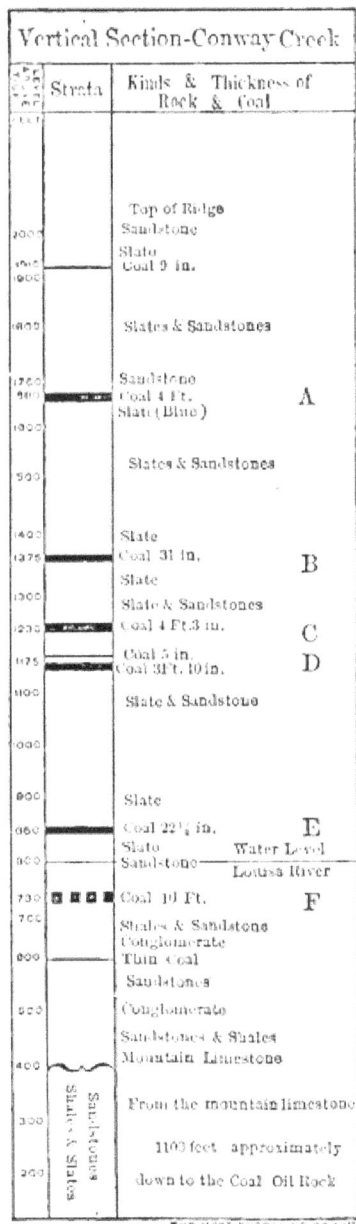

greater part of the county, and the rock material being uniform in character, the county is marked by a topography which is also uniform in its character. The elements in the course of time have imprinted the same features over the whole area, leaving deep hollows at intervals rarely ever exceeding two miles, in every one of which may be found more or less water, wholesome and pure. These hollows have sufficiently sloping sides to permit of the adoption of lines of roadway by side cutting of any grade desirable, or the location of shutes. The tops of the hills generally range from 800 to 1,000 feet above the main longitudinal section of the streams. Referring to the cross section, the coal veins there lettered will yield approximately as follows:

Coal vein A, like the others, a bituminous coking coal, with such a small percentage of sulphur as not to be noticed. Ash not over 4 per cent., and fixed carbon over 70 per cent.; will yield per acre 6,780 tons of 2,240 pounds per ton.

Coal vein B, 4,380 tons per acre.

"   C, about 7,200 tons per acre.

"   D,   "   6,500   "   "

Aggregating nearly 25,000 tons per acre.

Veins E and F are hardly necessary to calculate, one being thin, and the other, though thick, is below water level.

In the southern part of the county the veins measuring 4 feet thick are near the crests of the ridges. On Beech Branch, a small tributary of Lick Creek, and near Noah Countz's, at an elevation of 1,765 feet above sea level, one of the veins has the following dimensions, accompanying rocks, etc.: 58 inches thick, with 4 inches of slate parting near the middle; floor of slate; roof of slate 5 feet thick; then above this 4 inches of slaty coal; then 8 inches of coal; then 18 inches of slate; then 13 inches of sandy iron ore overlaid with sandstone.

Such readings could be repeated with but slight variations in many parts of the county. The dip is usually gentle in any direction.

SALT.

The salt-bearing series may be regarded as one of the valuable features, and the brine is likely to be reached in less than 500 feet below the level of Louisa or Russell's Fork, wherever any basin may be found in which the rocks dip toward a common center. Such a basin exists at Sand Lick, near the junction of Lick Creek with Russell's Fork of Sandy River. In fact much salt has been made there by boiling the water caught at one of the brine seeps in sunken barrels and hogsheads.

BUILDING STONES.

The building stones of Buchanan are in endless quantity, confined to a variety of sandstone found all over the county in several ledges. It is a well-known fact among the citizens of the county, that they can open a quarry a few hundred yards from their homes in almost any of the hills, in which they can obtain stones of great size, so soft when first exposed as to admit of being hewn into any desirable shape, even with a common axe.

TIMBER.

Buchanan has all the different varieties of timber known to this latitude, not only in great quantity, but the growth is usually of the best character. The girth of poplar trees frequently measures from 13 to 18 feet; ash trees, 7 feet, and all others proportionately large. To say that the trunks of the poplars are frequently 60 feet without a limb would be very likely under the mark than above it. This tree seems to be in its most congenial latitude here, and not only attains a fine size, but is more plentiful than any other except the beech and white oak. There is a great deal of walnut, linn, buckeye, and sugar tree; and the growth is so uniform over the county that one locality can scarcely claim any superiority over another.

Along and near to the larger water-courses a good deal of poplar timber has been cut and rafted down to the Ohio River; but, as compared to the whole, it is really a small percentage. There are yet remaining immense areas of virgin forest where nothing has been cut. In many places the walnut trees are very thick.

WATER POWER.

As has been remarked before, the streams are not constant through the summer; consequently reliable water powers

are scarce. Russell's Fork near "The Breaks" would give a
good and constant power, but nowhere else in the county
could a large volume of water be had in dry weather. The
streams, though they get very low, seem never to go dry;
and in the deep pools you find along their courses there are
yet quantities of fine fish—black bass, redeye, pike, etc.

AGRICULTURE.

Though the county is essentially a mountain plateau, cut
and seamed in every direction by the cañons in which flow
the streams, the steep land is usually fertile, very much so
indeed for sandy land. It produces corn and oats well.
Wheat and rye have a fair average yield, and sweet potatoes
seem to be in their native clime.

There are good-sized herds of cattle turned out annually
to range in the woods, where they do well. The whole area
of the county would suit admirably for sheep-growing or for
the Angora goat.

SCENERY.

Occasionally the scenery is very fine. A sketch of what
is familiarly known as "The Towers," on Russell's Fork of
Sandy River, has been attempted; but no one could give
even a fair idea on paper of the beautiful scene presented by
these high cliffs and rocks as they rear their lofty crests
nearly 600 feet above the stream below. It is a wild place,
still inhabited by an occasional wolf, perhaps the last of his
race in that quarter.

The fruits of Buchanan are generally more certain annually than in some of her sister counties. The peach tree
thrives well, and is a crop of great importance in the county.
Ginseng, though one of the herbs, has been in the past a
source of considerable revenue to the people of the county.

It is however disappearing under the constant drain which has been stimulated by the high price paid for it.

Grapes do very well in Buchanan; *bees* are quite a success, and *fish* are there in good quantity and of choice varieties.

TRADE IN CATTLE AND SHEEP.

Cattle..........1,200 head of stock cattle annually.
Sheep..........2,500   "

LINES OF TRANSPORTATION.

The proposed line of the Virginia, Kentucky and Ohio Railroad, which is to connect with the Big Sandy Railroad, passes down Louisa Fork, and will thus pass nearly through the center of the county from south to north. A branch of the Richmond and Southwestern Railway may pass through the county at some day.

TOWNS AND VILLAGES.

There are only few trading posts of note in the county except Grundy, the county site—Needmore, Shacks's Mills, Rock Lick, Knox Creek, Sand Lick, Grassy Creek, Dismal, and a few other places; but there are no towns in the county.

Grundy, the county site, has, besides the court-house, hotels, stores, and a weekly newspaper, which takes great interest in the advancement of the material welfare of the county.

PUBLIC SCHOOLS.

The public schools are being more carefully looked after by the authorities, and will, with a more dense population to sustain them, become an important factor in the welfare of the county.

GRIST-MILL, UPPER WATERS OF SANDY RIVER, BUCHANAN CO., VA.

(P. 254.)

FLOYD COUNTY.

This county, though comparatively new, is not far, if at all, behind its sisters in the importance of its resources. Any railway located with due regard to the position of its mineral veins and deposits must cause Floyd County to assume a position that would not only give it a high rank among the best counties of the State, but the character of the resources which would be thus developed would secure for Floyd a fame as deserving as it would be universal.

Like many other counties of this section, volumes could be written upon any of them before a satisfactory description would be reached; and it is a matter of deep regret that no more space can now be spared in the description of Floyd than this book is able to devote to it. Sufficient, it may be hoped, though, to show such prominent features of interest as will indicate its fine character.

This county lies between Montgomery and Pulaski on the north, and the counties of Patrick and Franklin on the south and southeast, almost touching Roanoke County on the east, and bounded by the rich mineral county of Carroll on the west.

Its northern line is marked in great part by a short section of Little River, and by Laurel Ridge and Bent Mountain. Its entire southern and eastern boundary is supposed to follow the crest of the southern bifurcation of the Blue Ridge, and its western limit has no very marked feature near it except "The Buffalo," a bold and beautiful peak rising many feet higher than the neighboring chains of mountains.

This county is watered by the head-waters of Roanoke River, by Little River, and some of the streams which flow west into Big Reed Island Creek.

The Geology of Floyd County may be described as lying between the Lower Laurentian Rocks, on the south, and the

Section through Floyd County.

Middle and Upper Huronian on the north, as may be more clearly comprehended by referring to the cross section on the opposite page, fifteen miles in length.

The whole series of strata has a trend between northeast and east-northeast.

Beginning in the Blue Ridge, on the south, there is a measure of pure *asbestos* eight inches or more in thickness, as at Barton's, between walls of steatite, which extend for many miles through the country. Next, northward, are valuable ledges of *steatite;* and about two miles farther north is the stratification which holds the valuable bedded veins of *copper ore* and *magnetite,* showing at the Toncray Mine (near the old Shelor Furnace), at the Bear Beds, at Weddell's, and other places.

This valuable *copper* vein, containing much arsenic also, of which it would be appropriate to speak just here, shows outcroppings at intervals for about fourteen miles in the county of Floyd. At the Toncray Mine it has been better developed than at any other point; and, judging from the more abundant show of rich gossan on the surface, it may be regarded as one of the best localities likely to be found on the whole length.

Dr. RICHARD O. CURRY, A.M., M.D., examined Toncray Mine just at the time when all the excavations were freshly made. In his very clear general description, he says he found the gossan (or Iron Cap) to be 30 feet deep from the surface, before the undecomposed sulphurets set in. This iron ore, says the Doctor, was used very extensively, and with great satisfaction, by the Shelor Furnace. No doubt the high character attributed to the metal then made by that furnace was owing to the use of the pure hematites and magnetites of this vein.

Dr. CURRY, continuing his account, says: "There are two tunnels driven in upon the vein, situated upon the declivity

17

of the ridge. The lower tunnel is driven in from the north side, south 40° east, so as to cross the lead, which has a course north 54° east. This tunnel reaches 245 feet, through a hard gneiss rock with quartz veins. Through the crevices of the quartz there are found small clusters of native copper. The main object had in view in excavating this tunnel was to obtain a drift for the upper gallery, expecting too that it would intersect the vein at a lower depth.

"The upper tunnel is situated about 70 feet above the lower, and has been driven in through gossan and vein rock to a depth of 300 feet. When this upper tunnel was first opened, it was injudiciously driven in too far to the left of the vein; but in carrying in a cross-cut to the right, about 40 feet from the entrance, the vein was reached at a distance of only 20 feet. It was then followed for 250 feet through a soft talco-mica slate, several cross-cuts being run off to the right and left, so as to test the width of the vein. Throughout this whole length the vein is traced without any intermission—increasing in richness and width as the depth descends. The cross-cuts and the tunnels, driven in parallel to the main drift, expose the vein in a most beautiful manner—the intervening partitions, which have never yet been stoped out, consisting of solid banks of ore, in all its varieties, but mostly the oxides and black sulphuret. As only 32 tons of ore have been shipped from this mine since it was first opened, and as the ores exist in such rich abundance all along its walls and roofs, it may be readily inferred that the company had but one object in view—to open their mine to its fullest extent before raising their ores. Consequently, they have been content to drive a tunnel of 6 feet width through a 30-foot vein, only bringing out such ore as they had necessarily to excavate in driving forward the tunnel. Thus they had exposed sometimes the center of the vein; then by a cross-cut they have run to its northern side; then,

by another, to the southern; and from each of these branches
carrying along tunnels parallel with the main trunk. They
have thus exposed this vein for 300 feet, proving it to be one
of great depth, with a width of 30 feet, the dip being to the
southeast, at an angle of 45 degrees. The average per cent.
of the ores raised is found to be 16. As soon as the lower
tunnel is completed, and has effectually drained the upper,
there will be no limit to the ores which may be excavated
from its tunnels and chambers. In many of the mines these
chambers are already formed by the continual raising of
ores; here, however, the intervening partitions between the
tunnels yet remain, and will afford, by stoping, an incalcula-
ble supply of rich ore."

At the Bear Beds, on this same lead, the *magnetite* occupies
a position in the northern wall of the vein about four feet in
width. It may not extend lower than the depth of the decomposed ores. Below that it may confidently be expected
that iron and copper sulphurets form the whole of the vein.

The dip of 45 ascribed to this vein at the Taneray Mine is,
in all probability, local, and confined to a short distance below the surface. The general dip of all the rocks of that
vicinity is much steeper.

This vein is also exposed at some old openings at the Hylton
Mine, Nowlin's, Howell's, and at other places, presenting the
same general characteristics, and yielding a very large tonnage of iron oxides, or gossan, on the surface.

It has been contended by some authorities that this iron
ore (the gossan) contains too much copper to make a good
welding metal. This is no argument against its availability
as a mixing ore, in a section containing so many varieties,
which, with proper railway facilities, can be brought together
in a less number of miles than can be done in any other part
of the country.

Next, north of the great copper vein, of value is a measure

of *soapstone*. From this extensive band many citizens of the neighboring country obtain supplies of building stone, which they can saw into shape, not only to be used as backing for fireplaces, but much other masonry is built of it besides, owing to the ease with which it can be shaped into the sizes desired.

This band seems to be but a repetition of another lying only a mile or so north of it, and might be taken for a repetition of the same stratum, caused by a fold. But, if this view of the case be correct, there ought also to be another exhibit of the copper vein. There is nothing, however, to suggest the probability of such a version, but a line of *magnetic ores* on the surface, showing at such places as at Whitelow's, Hogan's, about one and a half miles south of Jacksonville.

This line of magnetic ores, running northeast and southwest, has the reputation of showing in valuable quantities toward the eastern end of the county. Should the ore be found in sufficient quantities to be available, there is no evidence as yet to suggest the idea that they contain that objectionable constituent, titanic acid.

The *manganese* ore, outcropping here and there, in Floyd County, is not sufficiently developed yet to show its probable quantity and quality.

The next great mineral-bearing lead, encountered going north, is the galena-bearing quartz and the accompanying hydro-mica slates, etc., holding pyrites and decomposed ores in the shape of gossan. This band is over 200 feet wide, and shows quite conspicuously at Luster McAlexander's, on Little River. It is no doubt the northeastern prolongation of the Peach Bottom copper vein, which shows at several points farther southwest in Carroll, Grayson, Alleghany, and Ashe Counties. In the county of Floyd this lead can be traced for many miles. From its outcroppings, west of Abraham Burnet's, near Williams, through, northeastwardly, across Little

River and into the Beaverdam section, it can be found; but what commercial value can now be properly attached to it, it would be difficult to say.

At several points the quantity of galenite found in the quartz suggests the flattering hope of a sufficient percentage of silver to pay; but as yet it would only mislead public opinion to declare such a result attainable from present developments. Still, it is one of those vast depositories of mineral matter, which may at any moment reveal a great mass of highly valuable ore. It had more the appearance, a few months ago, of being valuable than the line of rocks from which the Brush Creek gold is supposed to be derived; and it would not be astonishing to hear of its being made the basis of very successful mining operations on an extensive scale. At a point near Winter's, on Terry's Creek, there are interesting masses of gneiss interstratified with talcose slate and some chlorite, containing pyrites of copper and iron and some galenite, disseminated in the rock.

In the immediate vicinity, about 1,000 feet north of the last-named vein, below McAlexander's, on Little River, there are plumbaginous and talcose slates. In these slates, in their eastern continuation toward the Locust Grove section, it may be that the *plumbago* is found which is, now and then, reported as occurring—near King's, on Bent Mountain Turnpike, and on Mills's lands, four miles from Copper Hill, for instance.

Passing thence northward, across great quartz veins, hydromica slates, etc., there is next encountered a series of kindred rocks containing the *gold* now being sought after on the east side of Little River, on Brush Creek, in Montgomery County.

GOLD.

On the Floyd County side of the river there is no stream answering in position to that of Brush Creek. Consequently the same formation, though containing, no doubt, the same percentage of free gold, is not so concentrated as to be notice- able in the same manner as it is where a stream, of the size of Brush Creek, has worn away so much of the rock material, leaving the gold in the sands and detritus along it.

It would be very interesting to know every feature of this important series of rocks ; but it is plain that to attempt more than call attention to the fact of the existence of gold in con- siderable quantities in these rocks, would be to enter upon an undertaking, which, to be properly done, would consume every page of the book.

It seems, however, that what gold there is, is, generally, very uniformly distributed through the great mass; but there is no reason to doubt that it is concentrated here and there in easily defined veins.

These Huronian rocks are known to have such character- istics in other places. Besides the gold, there is galena, and, now and then, copper pyrites with a still greater proportion of iron pyrites.

This gold-bearing series is then bounded on the north by about 3,000 feet of felspathic and quartzose rock interstrati- fied with impure steatite, hydro-mica slates, and sometimes talcose and chloritic schists, with hornblende almost totally absent. These rocks seldom assume the proportions of true granite. There are sometimes protogine, but much of the rock is an albite and orthoclose petrosilex. This, then, is the last series of the cross section on the north, in the county of Floyd.

In the northeastern part of the county, nearly in the con- tinuation, that way, of the gold-bearing series, on the North Fork of Roanoke River, there is a vein of magnetic pyrites near the house of Mr. Light, 12 feet thick, course north 45° east, dip 45° southeast.

This ore contains a great deal of copper pyrites disseminated through it, the iron pyrites being distinctly magnetic, and having the appearance of containing *nickel*.

It is somewhat singular that this vein should lie within twelve miles of the Atlantic, Mississippi and Ohio Railroad so long, with no up-grade intervening, without receiving more notice than it has. Also, at Purgatory, two and a half miles south of the north boundary line, PROF. FONTAINE's analysis of arsenical pyrites, found in a 5 to 7 foot vein, shows 18 ounces of silver to the ton.

It would be idle to deny the great value of Floyd as a mineral county. In nearly every part of it the surface indicates the presence of the ores described above. C. M. STIGELMAN, M.D., a gentleman of fine attainments in scientific pursuits, now residing at Jacksonville, enumerates some of the minerals and their localities as follows: Magnetic iron ore, at Whitelow's, Hogan's, Bishop's, Deskin's, Link's, and O'Connor's; micaceous iron ore, Runnet Bag Gap, right of Patrick Road, Blue Ridge; specular ore, eastern slope of Blue Ridge, Shooting Creek; argentiferous lead, Little River; arsenical pyrites, great copper vein; limestone, north of Court-House on Montgomery Turnpike; manganese, at Whitelow's and Columbus Rick's, Shooting Creek; asbestos, in the Blue Ridge, at Signer's and other places; pickeringite, three miles north of Court-House.

AGRICULTURAL FEATURES.

The proportion of lands in this county available for tobacco culture is, perhaps, greater than for any other purpose. In the west end, on the waters of Burnet's and Greasy Creeks, are the largest areas of grass lands; and upon these very considerable herds of cattle are grazed and wintered annually.

Throughout the county good farming lands are found,

though in many places very steep.   Occasionally a band of
talco-mica slates and schists, impregnated to a certain extent with manganese, will afford poor land where they form
the subsoil; but such areas constitute but a small proportion
of the whole.   Little River and its tributaries have many
fine farming tracts.   On Burk's Fork, though the land is generally steep, it yields a safe return to the farmers.   In fact,
much of it is really fine land, susceptible of a high state of
cultivation.   The rock here is frequently diorite, which, decomposing, has left a strong and permanent subsoil composed
of deep red clay.

Throughout the Blue Ridge the decomposition of much
gneissoid-rock material has left a soil which, though not
always of the first quality, is susceptible of improvement at
small outlay.   In the eastern end of the county, richer and
poorer lands alternate with each other, as the substratum is
more or less felspathic.   Much of it, judged by the growth
on it, is exceedingly fertile.   Down in the deep gorges made
by the North Fork of Roanoke River the lands are seldom
available for cultivation.

ANNUAL SHIPMENTS OVER AND ABOVE HOME CONSUMPTION OF
SEVERAL PRODUCTS.

Tobacco........185,000 pounds, mostly good wrap-
    pers, bringing $50 or $60 per hundred pounds.
Wheat.......... 11,000 bushels.
Corn...........   800   "
Cattle.......... 1,050 head @ $20 per head.
    "   500   "   "   8   "   "
Fine cattle......   100   "   " 25   "   "
Sheep 1,000   "   "   2   "   "

TIMBER.

Many sections of the county are still covered with a fine virgin forest. The Buffalo Ridge and Laurel Creek County is one unbroken forest for miles in extent. The spurs of the Blue Ridge, while rich enough to be farmed in many places to the top, are still heavily timbered with fine bodies of chestnut, chestnut-oak, white oak, etc. These trees constitute the main body of the timber in the more mountainous portions. There are valuable bodies of timbered lands, now and then showing a large quantity of white pine. This is mainly used in making shingles. Furnaces for making iron would not be at a loss, anywhere in the county, for a supply of cheap charcoal for some years to come.

WATER POWER.

Little River affords fine water. The Southwest Fork, near Jacksonville, is utilized to run one of the finest flouring mills in Virginia. In every part of the county, at intervals of a few miles, there are water powers, either in use as saw-mills, grist-mills, carding-machines, or only await the time when the increase of population in the county will require their use.

FRUIT.

Apples form one of the native fruits of the county, and rarely miss making a full crop annually.

Peaches are not much cultivated. Grapes, plums, and pears do well.

FISH CULTURE.

The streams are all well adapted to the culture of the finest varieties of game fish, but no attempt is being made as yet to stock any of them with improved varieties, except such work as has been done by the State Fish Commission, Col. Thomas

Lewis, of Salem, and Mr. Sumter, of Montgomery, in stocking New River and the Roanoke—some distance away from the county—the streams of Floyd being mainly tributaries of these rivers.

BEE CULTURE.

Many of the citizens pay a good deal of attention to the improvement of their bees, but none engage in producing honey for more than their home use. In this, no part of the country could be better adapted than Floyd. The number of flowering trees, shrubs, and plants, together with the numerous sucking places in the damp hollows, give the bee peculiar advantages. It seems that the bee is not only able to obtain from these marshy places the moisture it requires, but, in seasons when blossoms are scarce, must also derive some of the material used in making its honey.

TOWNS AND VILLAGES.

Jacksonville, where the court-house is situated, is the principal place in the county. It has churches of various denominations, hotels, and stores in which every variety of goods for country use may be had; a good and progressive newspaper, *The Floyd Reporter*, devoted to the advancement of the best interests of the community. It has good schools; and has such shops for repairs of wagons, tinware, harness, etc., as are usual in such towns.

PUBLIC SCHOOLS.

To judge from the tenor of the Report of the Superintendent of Public Instruction, the public schools have been neglected in the last few years, but there now seems to be a decided improvement in this particular. The schools will, no doubt, be more regularly sustained in the future.

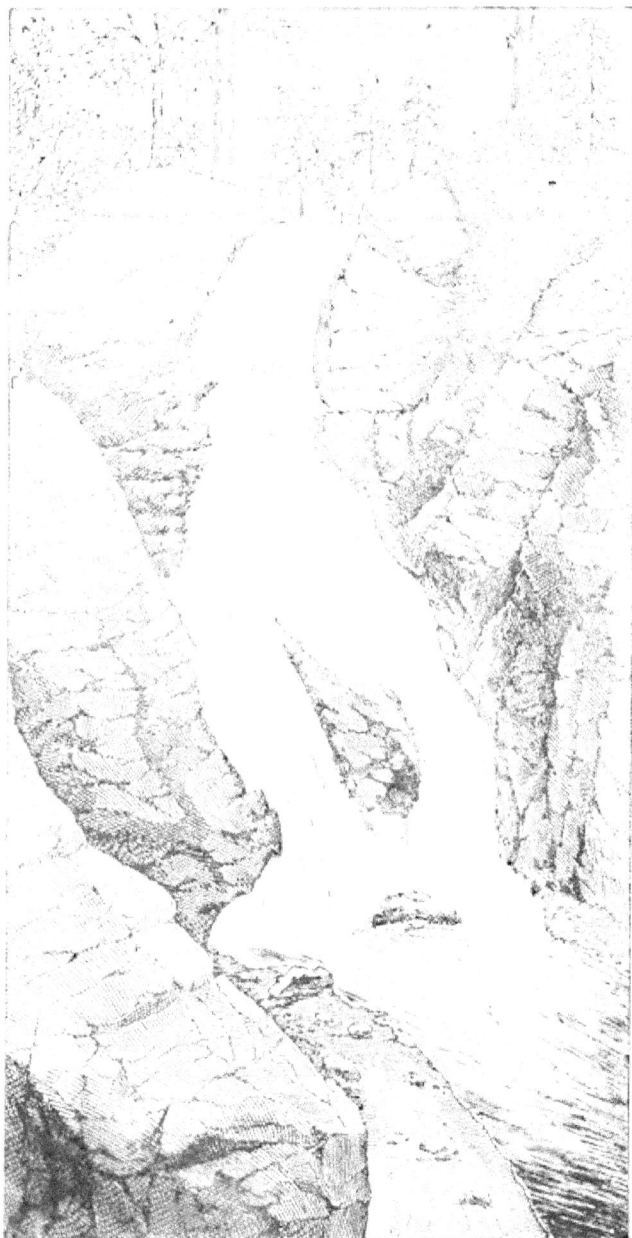

BEATRICE FALLS, FLOYD CO., VA.

(P. 267.)

SCENERY.

This county presents romantic scenery, of a kind beautiful and attractive, just in proportion, inversely, as it is unknown outside of its immediate neighborhood. The falls of the Roanoke, which are designated the Beatrice Falls, after the daughter of Queen Victoria—named more in honor of the fine personal character of those illustrious women than the positions they adorn—would compare with any piece of mere scenic beauty in the purity and harmony of all its surroundings. This fall of 90 feet, almost perpendicular, over a hard quartzose and felspathic rock, is terminated below by a pool, into which empties a smaller stream with an almost perpendicular fall of about 200 feet. This latter is called the Prince Imperial.

These two together form a rare picture, equaled, perhaps, by some of the scenery in California; and only excelled in point of magnitude by a few places, but not in beauty by any.

PUNCHEON RUN FALLS.

Only a few miles from the above-named falls, nearly on the line of Floyd and Montgomery, is another place of singular and attractive beauty. For about 350 feet, the water of a small creek, the Puncheon Run, dashes over the face of a great sheet of rock, with its sides fringed and hemmed in by every species of mountain vegetation. In the early summer it has a setting of red and pink and white, formed of the luxuriant blossoms of both the Rhododendrons, Catawbiense, and R. Maximum, the laurel, and the mountain ivy.

The place is wild and rugged, and, when better means of access can be secured, will form one of the most attractive features of this country.

All these falls are accessible from the Atlantic, Mississippi and Ohio Railroad, or the mineral springs of Montgomery County, in less than a day's drive.

CARROLL COUNTY.

This county scarcely needs any other introduction than an allusion to its character as the great copper county of South-western Virginia; a character which has been established, really, for years; but, until recently, through the publicity given to the facts, by the author of this work, in various lectures in the Eastern cities, and by the publication in Hotchkiss' able periodical (*The Virginias*), these vast deposits have remained comparatively unknown. It is true they may have been alluded to both by PROFESSOR ROGERS and PROFESSOR LESLEY; but at that early day there were so few developments as to afford only the most meager data upon which to base statements.

Carroll, being rich in other resources—in iron ores, timber, fine streams, mineral springs, etc.—must be regarded, each succeeding year that her unquestionably important resources are developed, as one of the most valuable of the brilliant gems that go to make up the remarkable series of mineral counties known as Southwestern Virginia.

In fact, so great is the body of sulphureted ores alone in the county, that they, with plentiful means of transportation, must form the basis of industries on a large scale, the extent of which it would be difficult, now, even to approximate. Thus, when the great West shall have exhausted the virgin strength of its soil, and becomes a much larger purchaser than now of good fertilizers, these heavily sulphureted ores will have been brought into easy communication with other valuable constituents—both those of South Carolina and of Carroll's sister counties—and will become, in the time of the country's greatest demand, one of the heaviest manufacturers of cheap and efficient fertilizers, perhaps, in the world. It would be curious and interesting to show how such industries could be established; what ingredients, such as potash, etc., could be brought together, and how they could be made into excellent and cheap fertilizers; but the fact that the lines of transportation are still wanting, as well as space, admonishes us to leave the subject for future consideration. In that time, no doubt, some cheap means will have been found by which those felspars of Grayson and Carroll, holding fourteen per cent. of potash, can be used in connection with the sulphur of the abundant Carroll County ores. It is true, this is some- what generalizing; but it will be only the uncandid mind which will be slow to admit the almost certainty of the above reflections, not to speak of the gigantic operations in the re- duction of copper even now seriously contemplated by some of the most experienced and capable men in that line of busi- ness in the country. That Carroll County, with adequate means of transportation, will develop mines of lasting and permanent value, there can be no doubt; and that this county will form one of the most considerable factors in the solution of the problem of the State's future prosperity, is beyond a question. It is able, by means of its vast hidden wealth, to bring lines of railway through the county, and will inevitably increase the tax-paying power of its own and surrounding communities to so great a degree as to render it a fit com- parison to say, that that capacity will have improved a thou- sand fold.

HOW BOUNDED.

Carroll is separated from Wythe and Pulaski on the north by the Iron Mountain range, locally termed here the Poplar Camp Mountains; northeast by Floyd County; southeast and southward it is divided from the county of Patrick by the main Blue Ridge; and touches the North Carolina line. Westerly it is bounded by Grayson County.

Section through Carroll County, Va.

HOW WATERED.

Carroll is watered by New River and some of its considerable tributaries, Big and Little Reed Island Creeks and their head-waters, Poplar Camp Creek, Crooked Creek, Chestnut Creek, and some minor streams. All of these being bold and regular in their flow throughout the year, give to the county a name for being well watered.

GEOLOGICAL.

The geological features of the county are nearly identical with those of Floyd and Grayson, except that Grayson has such an immense quantity of granite, which neither Carroll nor Floyd seem to have. The geology may, then, be said to be comprised between the Laurentian gneissoid series, near the heart of the Blue Ridge, and the Huronian, inclusive. The dip of the rocks has that general appearance of being monoclinal southwardly, or rather southeastwardly, common to most of the rocks of this region ; but in places there were once great folds or anticlinals, the crests of which have been denuded and swept away in the lapse of time since they were so folded, leaving both sides of the fold with the same general average inclination.

Beginning in the Blue Ridge, we have generally a gneissoid formation ; but, about the main crest, talco-mica, hornblende, and chlorite slates and schists and soapstone alternating with each other. In all these strata there are occasional heavy bands of quartz. The gneissoid formation prevails until you pass the general location of the Southern Copper Lode, as represented on the cross section. Then, in about half a mile north of this southern lode, you cross a broad band of hornblendic slates, schists, etc.; then soapstone ; then slate, schists, and quartz veins ; then, when within three miles of Hillsville,

on the south, you touch upon the trappean rocks in which is
situated the native copper lode; then, within a mile of Hillsville, alternations of hornblendic with mica slates and schists;
then about Hillsville, the continuation of the strata which
hold the northeastern extension of the Peach Bottom Copper
Lode, here mostly gneissoid; then, for more than a mile, hornblende; then, for three miles, going northwardly across the
great Northern Copper Lode and its branch vein the Dalton,
through talco-mica slates, chlorite slates, and sometimes
slightly hornblendic slates; then across several miles of a
repetition of hydro-mica slates, etc., to the foot of Poplar
Camp or Iron Mountain; then through the heavy quartzose
and slightly felspathic bands composing the mountain on the
south side, ending the section at the county line on the crest
of the mountain.

Following the plan adopted for the other counties, an attempt will be made to describe the iron ores first, although
the importance of the copper ores, comparatively, would
rather suggest the propriety of their being treated first. The
Iron Ores of Carroll, it may be submitted, cannot be regarded
as the least important of its resources. Not only do they
exist in very great quantity, but they are generally pure, except where they retain a little too much copper. It will be
understood that the greater part of these ores are derived
from the decomposition of iron and copper pyrites; even the
magnetites and semi-magnetites may be thus derived; hence
it is not to be wondered at that copper will be found in the
iron ores.

Of the *Brown Ores* there are vast beds and deposits existing as gossan along the outcrops of the different pyritous
veins, or lodes, of copper and iron. It would be difficult to
estimate the quantity which occurs on the southern or Ore
Knob Toncray Lode, as its greater distance from railway
transportation has caused it to be less explored than the

more northern veins. A description of the location of this lode just here would rob it of that interest which would attach to it in its character of a copper lode ; but it is not deemed to have that character for copper in Carroll County which it possesses either at Ore Knob Mine on the southwest, or at the Toncray Mine in Floyd County on the northeast. It is, so far as known in Carroll County, of more value in its character as an iron vein. This may be applied both to its massive exhibits of gossan, or brown and red oxide, here and there, and to the masses of undecomposed sulphurets existing below. This is observable where this lode crosses Snake Creek. The vein here exposed had not been so fully opened as to give a satisfactory showing of its true character when visited by DR. CURRY in 1859, since which time no work of consequence has been done in the way of development on the lode ; but enough has been done, and the outcroppings are sufficiently abundant throughout, to make evident the vast quantity of both gossan and sulphuret it is capable of yielding.

It is nearly twenty-seven miles long in Carroll, with a variable thickness between twelve and twenty feet thick, lying in the gneissoid system, just north of the Blue Ridge, dipping southwardly, generally at high angles. It has the interesting feature of uniting with the Native Copper Lode somewhere near the head of Laurel Creek, in the boundary line between Carroll and Floyd. Of this fact, however, the writer is not positively aware, as it was out of his power to follow up the Native Lode to the supposed junction ; and it is a matter of regret, also, that he was not able to give the Southern Lode as thorough an inspection in Carroll as he gave it in Floyd and in Ashe, or that he has given to the Great Northern Lode, the Peach Bottom, and the Native Lodes.

The next great bodies of *Brown Iron Ores* are found on the Great Northern Lode and its branch veins in the northern

18

central portion of the county. The vast gossan outcrops of
this lode, which passes, near Cranberry Plains, from southwest to northeast through the county, have been mentioned
by Dr. F. A. GENTH in his report to the "Wistar Copper
Mining Company" in 1876; by Dr. DICKESON in his report to
"The Dalton Mining Company" in 1857; by Dr. CURRY in
his "Visit to the Virginia Copper Region" in 1859, recently
quoted by Hotchkiss's "Virginias," and were described by
the author in several lectures delivered before meetings of
"The American Institute of Mining Engineers." Beginning
at the southwest end, bodies of this gossan, or hydrated per-
oxide of iron, are noticeable as the lode, after crossing New
River for the last time, passes into the county from Grayson,
about three miles north of Old Town. Here, near the Leonard Mine, in the Clifton Copper and Silver Mine, the lode begins to show those surface brown ores of iron which assume
such vast proportions a mile or two northeast at the Great
Outburst. It is possible that at the Great Outburst, and on
the Chestnut Creek property of Wistar Copper Mining Company, the ore is more than 150 feet thick by an average of
30 to 35 feet. Dr. GENTH, in speaking of the 6,800 feet length
of the lode, which he examined in 1876, says: "Taking the
average width at 45 feet, and the depth of the limonite (gossan) workable as a valuable iron ore at 30 feet, and the
weight of one cubic foot of this limonite at 150 pounds, the
Chestnut Creek property contains at least a body of 586,000
tons of available iron ore, yielding about 50 per cent. of pig
metal."

Again, from Copperas Hill, on Crooked Creek, where the
great lode appears to divide into two great veins, going
northeastwardly, through all the old workings—on the lands
of the Wythe Lead and Zinc Mines (which also own copper
property), Vaughan, Ann Phipps, Wild Cat, Cranberry, Dalton mines, Ann Eliza, and Betty Baker mines—you find a ton-

nage of brown iron ores in the shape of gossan which will go up into the millions of tons.

These ores—as at the Betty Baker mines—often present the appearance of highly valuable *ochreous deposits*. And the value of the whole is now only a question of cheap transportation.

SPECULAR ORES.

The pure ores of this variety are not as yet found in very large quantities in Carroll, though five veins are strongly suspected in this series of rocks. Near Thompson's Mill, on a hill north of a small creek which runs into Little Reed Island Creek, there is specular ore combined with magnetite in a vein not yet fully developed, but supposed to be six feet thick.

MAGNETITE.

Magnetic iron ore is found in many localities in Carroll; it is found to follow a line just north of the Southern Lode; and again another series of outcroppings is observable both north and south of the strike of the Great Northern Lode, as well as in many other localities. Unfortunately a want of transportation has prevented the citizens from taking sufficient interest in these ores to have them developed.

The magnetite as showing in surface pieces is usually very good; and there isn't enough titanium—as rutile—showing on the surface to warrant the belief that it is heavily impregnated with that impurity. As has just now been said, magnetite exists with specular ore in a vein which crosses a hill not far north of Thompson's Mill. Should this vein be six feet thick, as suspected, the quantity of ore it will yield throughout its length above water level would be very great, the hills being usually 180 feet at their crests above water

in the creeks. Nearer the Great Northern Copper Lode, on the
northern side, as well as on the southern side, the ores picked
up are a purer magnetite; but no satisfactory data have
been collected yet as to the thickness of the deposits.

IRON PYRITES.

To speak of the iron pyrites fully, again anticipates the description of the great metalliferous lodes carrying copper,
which would seem more properly to belong to the chapter on
copper.

Iron pyrites in Carroll is found in many of the rock strata.
Nearly every quartz lead has more or less of it. It is the
great basic material of the Southern Copper Lode. It forms
the greater mass of the Northern and Dalton Lodes, besides
minor ones it would be tedious to mention. To form even
the most distant idea of its quantity, it is only necessary to
imagine a length of 5½ miles, by a thickness of 30 feet, of an
unknown but very great depth. Much of this, strictly, is
pyrrhotite or polarized pyrites. A proportion of it is also
arsenical: the quantity of *arsenic* may assume, at points, large
proportions. In all probability, the quantity of arsenic in
the Southern Lode is much greater than in any other, judging
from the constituents it shows in Floyd County.

The iron pyrites, under favorable conditions of cheap transportation, would become an important basis for large chemi-
cal works, including the manufacture of fertilizers on a large
scale. Much of it being above water level, it could be mined
with great facility.

COPPER ORE.

It is not taking too much for granted to say that the copper
ores of Carroll County have become already very widely
known.

Besides the humble efforts of the author, LIEUT. MAURY, DR. T. S. HUNT, DR. F. A. GENTH, and other gentlemen eminent in scientific pursuits have had something to say about these veins. To these may also be added the efforts of DR. CURRY and of DR. DICKESON. To DR. CURRY may be ascribed the first effort to map out these veins, and illustrate them and the general geology of the country with proper cross sections; and the errors which he may have committed, here and there, are more than compensated by the mass of really reliable information which he gave at so early a day as 1859.

Taking the Southern or Ore Knob Toncray Lode into consideration first, it derives more of its importance as a copper lode from the splendid showings at Ore Knob, on its southwest extension, and the Toncray Mine, on its northeast extension, than from anything known of it in Carroll County ; but it may be assumed that it must, at greater or less depths, at many places in Carroll, contain rich ores of copper. It is a matter of regret that its greater distance, along here, from railway transportation, has prevented it from being more thoroughly developed.

NATIVE COPPER.

Next, going northwardly, is encountered the remarkable lode of native copper, which is known to run for eighteen miles in a direction which cuts diagonally across the general strike of the other veins. It was stated in the author's paper recently read before the New York meeting of the American Institute of Mining Engineers, that this lode had a direction from northwest to southeast. This was putting it strongly, to show the difference between its course and that of the Great Northern Lode, which is from southwest to northeast. The true course of the Native Lode is more east and west. It is apparently perpendicular in attitude—6 feet of close, hard

schistose material, holding native copper, having walls of tre-
molitic and hornblendic trap, in some places yielding beryl.

Its location, as determined by actual survey, is shown on
the large map accompanying this book. It will be seen from
that, that it intersects the Peach Bottom Lode near Woodlawn, and traverses the formation in the direction of a point
where the Southern Lode crosses the boundary line between
Carroll and Floyd. It may be submitted that there, on the
head-waters of Laurel Creek, valuable discoveries may be anticipated.

No estimate has as yet been given of the percentage of copper in this vein above eight per cent. The vein can be seen
exposed at James Early's, south of Hillsville, and at the point
where it crosses Reed Island Creek, besides other places.

THE PEACH BOTTOM LODE.

This lode can be clearly traced from the southwest side of
the county, very distinctly, as far as Hillsville. It passes
close on the south of Woodlawn, and there on lands of that
vicinity it shows plainly in nearly perpendicular ledges
of micaceous schists interlaminated with quartz. These
rocks are gneissoid in structure as you approach Hillsville;
and there they are in such workable ledges that some of the
stones containing copper pyrites are now in the foundation of
the court-house. It will be as well to repeat here what is
said in the description of the Peach Bottom Copper Mine, that
the ore of this vein, everywhere it has been examined, is copper pyrites, with some carbonate of copper near the surface,
resulting from decomposition of pyrites, and, for the greater
length of the vein, silver-bearing galenite. It is almost impossible to give a correct idea of its thickness or value in
Carroll. So little has its presence been suspected by the
mining world that sufficient developments have not as yet
been made by which to judge correctly.

A *quartz lode* containing pyrites of iron and copper, in distinct crystals, of large size, is next found between the Peach Bottom Lode and the Northern Lode, lying nearer the latter than the former. It is not probable that it will prove of value, as the quartz forms too large a percentage of the measure.

Next north of this is the Great Northern Lode of copper and iron pyrites.

NORTHERN LODE.

It may be an error to call all of these metalliferous veins by the technical term lode. As to this one, it has the appearance just west of Chestnut Creek of occupying a perpendicular fissure in a dipping stratification; while at a point near Early's, at Cranberry Plains, it dips 45 about, in a direction southeastwardly. At the latter it would be regarded as distinctly a bedded vein ; at the former, a fissure vein or lode.

This great vein or lode, coming from the direction of Ducktown, makes its appearance in Carroll County about three miles north of Old Town on the Carroll-Grayson line, and extends, without intermission, twenty-five miles, northeastwardly, to the Carroll-Floyd line.

Its first southwestern opening of any consequence is at the Leonard opening, sometimes called " The Clifton Copper and Silver Mine," so named because of a flattering analysis returned by an Indianapolis chemist, in which he ascribed to this part of the vein the remarkable quality of holding (besides sulphur, iron, and copper), nickel, silver, and arsenic. A great many persons do not credit this analysis, which was made for Mr. EDWARD SHELLEY, of Wytheville ; but when the chemist was applied to, he persisted in saying that the ores sent him gave the results reported. It is somewhat singular that Dr. GENTH did not find silver or nickel in the ores about Chestnut Creek, two and a half miles farther northeast. MR.

DEAN, of Indianapolis, reported one mass of the ore sent him
to contain over $450 to the ton in silver ; but as the matter is
entirely too important to be left to any degree of conjecture,
no fuller report will here be given of the analysis, the hope
being entertained that selected samples of the ores may be
forwarded to different competent chemists for thorough assay.
Next to the Clifton Mine is the Great Outburst—a name applied to an immense surface exhibition of gossan—on the western boundary of the tract of the Wistar Copper Mining Co. ;
then going east less than a mile you are on the main mining
ground of this company, of which DR. F. A. GENTH made a
very close and exhaustive examination of 6,800 feet of the
lode near Chestnut Creek, in 1876, at which examination the
author was present.

Although DR. GENTH says the vein is "between 30 and over
60 feet," it is believed to be, at one place, over 150 feet thick.

DR. GENTH says : "The geological formation on these lands
is that of gneissic rocks, more especially consisting of mica
slates and schists, graduating into chloritic and talcose. Their
average dip is about 45° southeast, with a strike between
north 30° east and north 45° east.

"A very large mineral deposit between 45 and 65 feet in
width coincides in its course with the strike of the inclosing
rocks ; from the developments made by Shaft No. 1, the ore
deposit seems to intersect the rock-strata, and to be nearly
perpendicular, corresponding in this respect with similar deposits in the same range of mountains, for instance, with that
at Ore Knob, Ashe County, N. C.

"The ore deposit has been developed by numerous shafts
and several tunnels, many of which have been made long ago,
and are at present inaccessible.

"About 3,400 feet from the southwest boundary line of the
property, Shaft No. 1, or the *Pyrites Shaft*, has been sunk to a
depth of 105 feet altogether in the vein.   At the bottom of it

a tunnel has been started toward the southeast wall, which, however, was not reached.

"This shaft was started in the decomposed part of the vein, the so-called 'gossan,' a hydrous ferric oxide or limonite. At the depth of ten feet this was occasionally stained with green carbonate of copper or malachite, which sometimes occurred even in larger masses.

"At a depth of between 20 and 25 feet the limonite was penetrated, and a bed of black copper ore of about three feet in thickness was reached. This consisted chiefly of black oxide of copper, copper glance, and small quantities of copper pyrites. The analysis of samples of this ore yielded 21.08 and 28.90 per cent. of metallic copper.

"At a depth of 25 feet the oxidized ores had completely disappeared, and were replaced by the undecomposed sulphurets, mostly pyrrhotite or magnetic pyrites, with some iron pyrites and small quantities of copper pyrites.

"These developments show that the vein was divided by interlaminated talc into three seams; that near the hanging wall and foot wall containing a far smaller percentage of copper pyrites than the central seam, which latter is about 10 feet in thickness, the total thickness of the vein being here about 30 feet.

"Although not free from copper pyrites, the ores in the other seams consist mostly of pyrrhotite. The central seam, with a considerable admixture of copper pyrites, was struck by the shaft between 40 and 50 feet depth, and was again cut by the tunnel at 105 feet, showing the same character, but with a decided increase of copper pyrites at the lowest depth reached, from which it is safe to presume that this most valuable and reliable copper ore will, in greater depth, replace the leaner sulphurets of iron.

"As in the developments made by this shaft the ores had not been kept separate, I have selected with great care, from the ores on the surface, a sample representing the lowest yield of the same. It was found to contain 1.70 per cent. of copper; ores from the central seam and the southeast tunnel yielded 9.36 per cent. of copper. It would have been easy, if this had been desirable, to select ores yielding from 30 to 35 per cent. of copper.

"Near this shaft, on the west side of the vein, a tunnel was started, by which developments, however, no additional information could be gained.

"Ascending the hill toward northeast, the Shaft No. 2, or *Whim Shaft*, is reached at a distance of 1,500 feet. This shaft was inaccessible, being full of water. Ores which had been raised from it, and which were lying on the surface, showed the character of the deposit.

"After the penetration of the gossan or limonite, rich black ores were found at a depth of about 40 feet, in a layer of about 14 inches in thickness, accompanied by about $2\frac{1}{2}$ feet of leaner ores, consisting of pyrrhotite, copper pyrites, a little black oxide of copper associated with talc and actinolite, and yielding about 10 per cent. of copper. A sample of the black ore from this shaft yielded by analysis 51.53 per cent. of copper.

"This shaft has not been sunken deep enough to reach to undecomposed sulphurets.

"Here the vein has been stripped across for 45 feet, but had not reached the walls.

"At a distance of 414 feet from Shaft No. 2 is another one, and 24 feet further on a fourth, and 96 feet northeast from this a fifth, the latter of a depth of 80 feet. They are all inaccessible, and in part caved in.

"There are several other shafts and tunnels on the 1,350 feet of vein between the last-named shaft and the northeast boundary line, neither of which could be examined on account of their inaccessibility.

"By these developments the character of the ore deposit has been shown to be the *same throughout ;* large quantities of rich black copper ores have been removed by these workings.

"There is another point of importance to which I wish to call your attention, namely, to another copper deposit, which lies southeast of the main ore deposit at a distance of about 80 feet.

"It has been proved by a shaft and tunnel, No. 3, *not* to be connected with the main vein. I could see in the tunnel only a small part of that which was left from previous workings, which was a body of black ore about 18 inches thick and 15 feet in length. A sample, apparently representing a fair average of the ore, yielded by assay 10.24 per cent. of copper.

"It is not improbable that this deposit originates from the *reprecipitation* of the copper, leached out, when the upper part of the main vein was decomposed and converted into limonite, and it may extend the whole distance of the main vein and parallel with it.

"I consider it of great importance to make developments to ascertain the value of this suggestion, as the ores from this deposit can be easily mined.

"Summing up the observations which have been made on the main vein, we find the following data:

"The length of the vein through the Chestnut Creek property is proved for a distance of over 6,800 feet. The vein can be traced over the whole property by a very bold outcrop of limonite or so-called gossan. The thickness of the vein is between 30 and over 60 feet.

"According to the elevation, the oxidized ores, mainly limonite, free from sulphurets and with traces only of copper ores, form the upper part of the vein to a depth of from 20 to 40 feet.

"Below the limonite is invariably found a rich layer of copper ores from one to three feet in thickness, and yielding from ten to over fifty per cent. of copper.

"Below this occur the undecomposed sulphurets, principally pyrrhotite or magnetic pyrites, intermixed with small quantities of copper pyrites, the latter increasing as a greater depth is reached, from which fact it is reasonable to suppose that this ore may soon be found in paying quantities.

"Based upon the above data, the following will be seen to be a low estimate of the valuable ores at present existing in the main ore deposit of the Chestnut Creek property, leaving out of consideration the sulphureted ores.

"Taking the black ores only at one foot in thickness, there are over 300,000 cubic feet of copper ore, representing about 20,000 tons of ore, yielding not less than 4,000 tons of fine copper.

"From these data, and the fact that the undecomposed sulphurets show a decided increase in the yield of copper pyrites in depth, the great value of your property is self-evident."

Then, passing on northeastwardly through numerous good properties along the lode, such as the old Limeberry, Copperas Hill, Wythe Lead and Zinc Company's copper lands, Vaughan's, and others, we come to the old J. EARLY property, now owned by the Baltimore firm of CLAYTON & WILLIAMS, as well as being owners of about nine miles' length on the lode either way from this property. When DR. CURRY visited this property in 1859, the shafts and tunnels were then in better condition for exploring the vein than they have ever been since. In fact, little or no work has been done since then of a reliable nature; that is to say, very intelligently directed.

DR. CURRY says: "The property, extending one half mile on the lead, was opened in 1854, and the work pushed to a greater extent than on any other property. This property is composed of an elevated ridge, which rises like a crest over-looking the Wytheville Turnpike, and well adapted to the tunneling to which it has been subjected. The total drivage of levels amounts to 800 feet, opening upon a mineral vein about 60 feet below the surface, and running north 54° east, with a dip of 60° to southeast. The entire vein, in all its length through the property, is estimated at not less than 10 feet thick and 25 feet wide. There are about nine shafts sunk on the lead, for ventilating mainly, their total depth being 250 feet, though the deepest only reaches 45 feet. Neither the depth nor the width of the mineral vein has been fully ascertained. Cross-cuts have been made from the main tunnels, and parallel levels driven, but still along the mineral lode. The works on this property exhibit very markedly the order of superposition of the various ores of these mines. After penetrating through the gossan crust, which here is strongly deposited, the carbonates and oxides are found occupying the upper portions of the veins. To these succeed, in the second galleries, the decomposed bisulphurets or black ores ; and, in the lower gallery, the gray and blue bisulphurets, beneath which lies the mundic rock. These galleries are separated by thin floors of rock, or of plank, and beautifully illustrate the system of mining (1850) in following the vein downward. A deep shaft has been also sunk in the valley at the base of the ridge, and near the turnpike, which, after passing through a hard, quartzose slate, opened upon a vein of the yellow sulphuret. This shaft, in our opinion, would have yielded handsome results, had it been located a few paces farther to the south. It also establishes the fact, that below the mundic comes the yellow sulphuret in the vein rock, which would grow richer as the depth increased.

"There have been 700 tons of ore shipped from this mine, consisting of the usual varieties of carbonates, oxides, and sulphurets, and averaging 15 per cent."

Going, then, northeastwardly, over several valuable openings on the lode, the Betty Baker Mine is reached, of which Dr. Curry says: "Entering the levels opposite the Anna Mary Mines, we pass in for 40 feet at right angles to the lode, where it is reached. It is then followed 300 feet, with some two or three cross-cuts and parallel drifts, exposing, throughout its entire length, a splendid view of the vein, from which the red and black oxides are mined, yielding 22 per cent. Since the 1st of May last, 130 tons of 20 per cent. ore have been taken out and shipped to Baltimore. Here, as in the Cranberry Mines, the richer ores occupy the upper vein, while the poorer lie upon the mundic rock, beneath which no explorations have been made (1859)."

Since that time (1859), however, very extensive explorations have been made by Mr. James E. Clayton, of Ore Knob, and the yellow sulphuret found below, as is usually anticipated in veins which show such a quantity of decomposed ores nearer the surface.

It may be remembered, in speaking of iron pyrites, it was stated that this Northern Lode seemed, in going northeastwardly, to divide at Copperas Hill, and to present from there two veins. The southern of these two is called the Dalton vein. In fact, during the time when copper was commanding a high figure in market, previous to 1861, the excitement here was so high that everything containing the slightest trace of copper was magnified into a vein or lode; hence, we are informed, by some of the old reports still in existence, that in the vicinity of Cranberry Plains there were three distinct veins, known as the Early, Dalton, and Dickeson leads. Dr. Dickeson, then of Philadelphia (1857), in reporting to the "Dalton Mining Company," says:

"The geology of this mining property belongs to a somewhat complicated series of rocks. The lodes are contained between walls of talco-micaceous slate, belonging to the Silurian epoch or period, but the metamorphic influence they have been subjected to has greatly modified its character. The summits of the hills we find composed of primordial rocks, consisting of imperfect granite, mica slates, talcose slates, and immense quartz rocks. On the western descent we meet with a series of shaley sandstones and slates, all of a metamorphic character. Descending into the valley and ravines, the rocks partake of the graywacke and conglomerate series, with alternating layers, varying in composition and color from an ashy gray to a pale blue tint.

"On the southwestern slope the gossan outcrop is very remarkable for its bright red color, and the disintegration, caused by atmospheric action, produces a beautiful and permanent pigment, that might be applied to many useful purposes. Two large quartz veins occur on the property, and in places large masses are scattered about in great confusion, completely intermixed with the gossans. Three well-defined metalliferous leads coursing north 2¼ east, and nearly parallel to each other, may be traced upon the surface by the gossan outcrop; and, by the depression of the exposed strata, the angular dip was ascertained to be 35 .

"Along the sides of the ravines numerous prospecting openings have been made, and each, as far as the character of the lode is concerned, shows the same deposit of copper ore. There are two large veins of quartz on the Dalton property, which I have found to be identical with the matrix or gangue stone of the yellow sulphurets exposed in other workings upon this lead. This rock, by comparison with that of the east side of these leads, differs in not being liable to the same decomposition by exposure. This gangue I have traced for several miles above ground, and it seems to lie contiguous to the iron lead (Great Northern Lode), which follows nearly the course of the mountainous ridge.

"The average width of the quartz veins is about five feet,

and, if we should include the numerous ramifying branches that set off from innumerable points, it might be stated much wider. Farther on, we find a coalescence of all these ramifications; and, if we are to take the mining rule for granted, there must exist, at no great distance, a heavy deposit of ore.

"The Dalton Mine is situated upon the lead of the same name, and consists of seven regular shafts, sunk about 35 feet in depth, cutting the lode of smut ore, which enveloped considerable bodies of black, gray, and red oxides of copper. From these shafts there have been driven four horizontal galleries or levels of variable lengths, from 40 to 150 feet, the lode in these dipping at an angle of 45° from the horizon.

"There are four shafts, which have lately been sunk under the direction of HENRY ANSELOTE, who now has charge of the workings. The southern shaft was sunk 42 feet, through a bed of light gossan, and at this depth it coppered over the entire shaft; north, from the above-named shaft, at the depth of 36 feet, a fine body of copper ore was cut. They sank on the north wall, drove east, and exposed an immense lead of copper ore. Four hundred yards north a prospecting shaft was sunk 25 feet through gossan, and at that depth struck the lode; this shaft was 40 feet off the lead; beyond this, some 250 yards northeast, they cut through 25 feet of gossan and struck copper ore; but here the water came upon them and drove them out.

"A shaft has also been sunk on the large quartz vein to the depth of 40 feet, where numerous nests and bunches of the yellow sulphuret of copper were found, in every respect resembling that of the Fentres and M'Culloch Mines, Gilford County, N. C. From the observed course of this pyriteous vein, I found it to be that of north 24° east, with a dip that seemed strongly tending to verticality. It is very desirable that these veins should have a vertical dip, as all mining operations are much more simple on erect than flat veins, for there is less cross-cutting required, and fewer winzes have to be sunk in the levels. Seams of white quartz, interlaid with seams of chloritic green-stone, occur very often along the vein, forming small feeders, and invariably indicate a greater deposit of ore where they unite. This vein does not exactly conform to the gossan outcrop lying north of it, nor does it entirely agree with the other quartz veins in the vicinity ; and, from the appearance of the gangue in the 40-feet shaft, I should judge the sulphuret to be not far off.

"A TABLE OF ANALYSIS UPON SAMPLES OF ORE OBTAINED FROM THE DIFFERENT SHAFTS AND OPENINGS UPON THE PROPERTY OF THIS COMPANY.

Mines.	Copper Ores.	Percentage.
Dalton........	Red oxide............	63.04
"	Black oxide...........	54.02
"	Smut................	25.12 "

It may be possible that enough has been thus shown to prove the inexhaustible richness of these metalliferous veins, under a proper system of mining, aided by cheap transportation.

It would tax the patience of the reader too severely to give more details of mining operations on the Great Northern Lode ; but enough work has been done, along its whole length, to prove that it has great thickness—sometimes reaching 150 feet ; and that it is sufficiently charged with copper to render it one of the most noted copper veins in the world.

GOLD AND SILVER.

No well-ascertained information has been obtained, as yet, concerning gold and silver. Gold is strongly suspected in the Huronian series, occupying the southwestern prolongation

19

of the Brush Creek Rocks, which are now yielding gold in
Montgomery County. Silver is reported to exist in large
quantities in the ores of the Clifton Mine, in a vein about
eight feet thick ; but silver mining would seem incredulous.
The same may be said of *nickel* and *arsenic.*

MICA.

No mines, yielding large mica, have been opened yet in
Carroll, though large mica is known to exist in the southern
part of the county. *Asbestos* is also known to exist, but is not
yet explored.

BUILDING STONES.

The granitic and gneissoid rocks in the southern part of the
county afford fine building materials ; some of the granite has
been used with satisfaction for millstones.

About Hillsville fine ledges of gneiss afford good building
stone.   The soapstone ledges are also called into requisition,
from which to make linings to fireplaces, jambs, etc.

MINERAL SPRINGS.

The mineral waters of Carroll have long been known
throughout this section as among the most curative in Virginia.

The old Grayson Sulphur Springs, on the north bank of
New River, twenty miles south of Wytheville, has four
springs, "all of which issue from a slate rock ; three of them
have openings near each other, within an area of thirty feet
in diameter, one of the springs, situated immediately under
bluffs, and at all times preserving its perfect transparency
and limpidness, in the small basin which has been excavated
around it, flows off through a channel; upon which, immediately after leaving its basin, it commences to deposit the
peculiar white material from which the characteristic title

of White Sulphur is derived. The next, which is similarly arranged, throws down a reddish-brown color, hence the name of Red Sulphur is given it. The other two appear to be chalybeate. The temperature of these springs, which appears to be between 47 and 48 degrees, is so low, that besides furnishing a cool and refreshing draught, they are able to contain their gaseous contents much longer in a state of combination, thus giving them a decided advantage. Owing to the presence of carbonic acid gas, the water is found to be what is called in familiar terms ' a light water ;' a term designated to express that several glasses may be taken without experiencing any sense of oppression."

These waters have the powerful adjuvants of wild and romantic scenery ; river and mountain combining to render a lovely scenery more attractive, and a fine and inconceivably bracing air giving a zest of enjoyment which cannot possibly be equaled. If that were not enough, the fine trout streams of the vicinity afford noble sport, and the hills still contribute a deer now and then for the chase.

TIMBER.

This county is well supplied with extensive forests of all the trees known to the latitude except fir. Toward the southern side of the county the timber areas are vast—almost unbroken. There are very good bodies of white pine in the northwestern section of the county, and in other portions. A great deal of the northern half of the county has been cut over, and the timber made into charcoal, to supply the iron furnaces and the lead mines in Wythe County ; but there is enough remaining to supply a large demand for years to come.

WATER POWER.

If the water power of Carroll could ever be utilized, there is enough of it surely to supply an almost unlimited requisi-

tion. The fall of New River, as it passes through the county,
being fully fifteen feet per mile in some places, would give a
power calculated upon a discharge of 1,150 cubic feet per
second in very low stages. Chestnut and Crooked Creeks discharge about 50 feet per second, and fall rapidly. Big Reed
Island Creek, discharging for a great part of its length 100 feet
per second and having a good average fall, would supply fine
powers ; Little Reed Island Creek, of about two thirds the volume of Big Reed Island Creek, offers many fine sites for mills,
etc. Poplar Camp Creek likewise offers a good many mill
sites, though it is much smaller than the above named. These
creeks, with their larger tributaries, supply every section of
the county with ample water facilities of every kind except
navigation.

MANUFACTURES.

At this time Carroll may boast of an iron forge or two, and
a few carding machines ; but beyond these its want of transportation has retarded its movements in the direction of
manufacturing. No doubt the writer of a few years hence
will have to record a large number of various kinds of works,
strewn along the river and the great metalliferous lodes, if
the well-known existence of great resources is any incentive.

AGRICULTURE.

The soil of Carroll, like that of Floyd County, is fertile
where the underlying rock strata do not partake too heavily
of manganiferous epidote and slate. All ledges partaking of
trap in their composition, when decomposed, leave a rich and
permanent soil ; hornblende schist usually follows the same
rule. Talco-micaceous rocks usually leave a soil easily
washed, which is not regarded for strength. The absence of the
great granitic masses, which mark the northern side of Gray-

son County, and make it so rich, leaves the northern part of Carroll not so well off in agricultural as in mineral wealth. The southern half of the county seems to possess by far the largest areas of strong lands, except where the lands are sufficiently level, as those near Woodlawn, to intercept all fertilizing material and hold it.

In many places in the county there are good grass lands, even to the crests of the high hills. Wheat usually does well, perhaps better than corn. Buckwheat, rye, oats, and potatoes make never-failing crops. Cattle, sheep, horses, and swine do well in the county, as a general thing, being raised at very small cost.

SCENERY.

In various parts of the county the scenery is wild and picturesque. About the river many beautiful views meet the eye. Throughout the county the number of wild and romantic dells, nearly always having cascades and waterfalls, is almost beyond belief. The county cannot be said to be lacking in this particular.

FRUITS.

Carroll, like Grayson, is a good apple county; but, as to other fruits, it is subject to the same risks, as to constancy of yield, that neighboring counties are.

Grapes could be made a specialty in the county with success.

Bees form an important feature, and thrive well.

Fish will soon be numerous in all the streams, as soon as the improved varieties recently placed in New River become more numerous. Speckled trout are quite common in the streams, proving the character of the streams for fine fish.

TRADE IN CATTLE, SHEEP, ETC.

Of cattle there are 3,000 head sold annually, of which 2,500 head are stock cattle, and 500 head are fat cattle. Many of them go to Danville and Salem for shipment.

There are 100,000 pounds of bacon annually sold at Winston, N. C., about 50,000 pounds to other points, 300 horses annually, and 50 mules. Of sheep there are about 3,000 sold annually, and 25,000 pounds of wool, some of which is consumed at home.

LINES OF TRANSPORTATION.

The U. S. Government has had the river examined with a view to its improvement; and the officer in charge has made a report favorable to the scheme.

The Altoona Coal and Iron Co., of Pulaski, are contemplating an extension of their narrow-gauge railway through the county.

The Pittsburg Southern Railway, if extended, may pass along the western side of the county, following the river.

Also, the New River Railroad, now being constructed between Hinton, Chesapeake and Ohio Railroad and New River Depot, Atlantic, Mississippi and Ohio Railroad, may be continued up the line of New River, through Carroll, on its way to North Carolina.

TOWNS AND VILLAGES.

Hillsville is the county site, nearly in the center of the county. It contains, besides the court-house, churches, hotels, stores, a good county newspaper, smith shops, repair shops, schools, etc.

There are numerous good trading posts in different parts of the county, but there are no places as large as Hillsville elsewhere in the county. Its manufacturing facilities will, no doubt, one day cause a great change in this particular.

Public schools will hereafter fare better in Carroll County; the principal obstacle heretofore has been the irregularity in the State appropriation to the schools; hereafter that point is met by stringent enactments of recent legislatures.

GRAYSON COUNTY.

While Southwestern Virginia can boast of two or three good counties located on the plateau of the Cumberland Mountains, carrying with them all that reputation for wealth in coal which the name Cumberland usually implies, she can point with equal pride to the counties lying on the Blue Ridge plateau, with their almost immeasurable wealth of copper ores, magnetic iron ore, gold, and that numerous list of valuable minerals nearly always found in the metamorphic series of rocks.

Grayson County occupies an enviable position in the list of Blue Ridge counties, albeit some of the geographers of the day try to throw the Alleghany Mountains to the south of this unquestionable Blue Ridge system.

Grayson, with its lofty, picturesque mountains, high waterfalls, beautiful rivers and streams, fine grazing and farming lands, ores and minerals and noble forests, is not second to any in the promise of a fine future.

It will be regarded as a fortunate circumstance if even partial justice can be done to Grayson in this work; but this need not be expected, for it is one of those subjects upon which a volume can be exhausted, and still the bulk of the story remain untold.

HOW BOUNDED.

Grayson is separated from Wythe, Smyth, and Washington counties on the north and northwest by the Iron Moun-

tain Range, which is the true southwestern prolongation of
the north-lying bifurcation of the Blue Ridge, the rocks being
identical with those described by Prof. Fontaine as marking
the Blue Ridge farther east in Virginia.   On the south side
is the State boundary line between Virginia and North Carolina ; east it is bounded by Carroll County, and west there is
a small length bordering with the State of Tennessee, being
separated from it by irregular ranges of lofty mountains
belonging to the Unaka system.   In this series, but not at
the corner of the States of Virginia, North Carolina, and
Tennessee, as has been supposed, is the well-known "White
Top," which rears its crest nearly 6,000 feet above sea level.

HOW WATERED.

Grayson is well watered by New River and tributaries.   It
gives rise to no other waters, except perhaps a few of the
head springs of Laurel Creek—a tributary of Holston River.
It may be considered one of the best-watered counties in the
State.   Not only does New River carry at all seasons sufficient water for even navigable purposes, if improved ; but its
numerous tributaries, flowing from never-failing springs, supply a wealth of fine water for every purpose.

GEOLOGICAL.

There is a difference of opinion among geologists as to the
exact classification of the rock formation in Grayson.   It may
be stated here with confidence, that the assertion will be
finally sustained, that the southern side of the county is in
the gneissoid system, belonging to the Laurentian rocks, and
that the northern side of the county is marked by the rocks
of the Huronian epoch.   The trend, or direction of the outcrop of the strata is between north 45' east and north
65' east.   The dip is usually southwardly, or rather, at

VERTICAL SCALE, LEVELS IN FEET ABOVE TIDE

Iron Mountain

REGION OF GOLD BEARING ROCKS

El^k Creek

White Top

Buck Mtn.

Pt. Lookout

Section through Grayson County, Va.

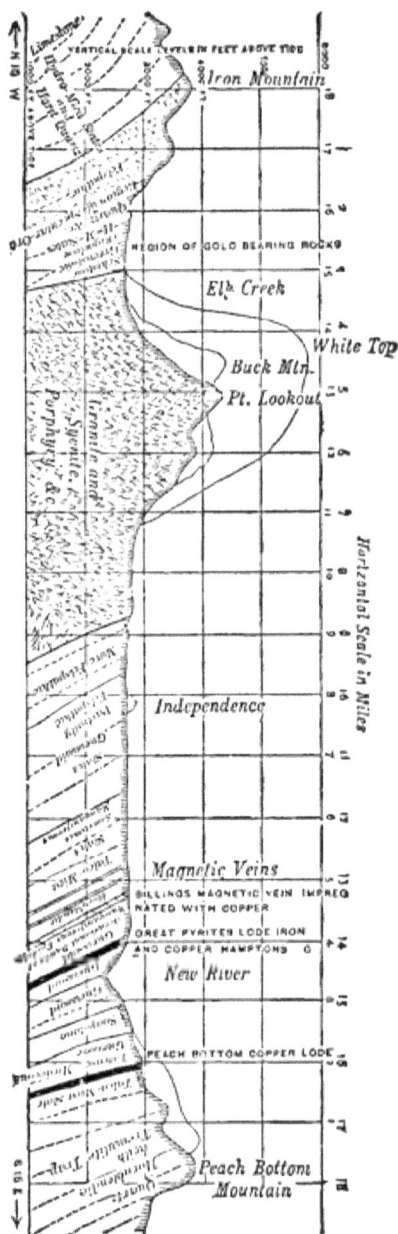

Syenite and Porphyry, &c.

Independence

Horizontal Scale in Miles

Magnetic Veins

BILLINGS MAGNETIC VEIN IMPREGNATED WITH COPPER

GREAT PYRITES LODE IRON AND COPPER HAMPTONS G

New River

PEACH BOTTOM COPPER LODE

Peach Bottom Mountain

13*

right angles to the trend. It is apparently monoclinal, that is, the ledges all seem to dip in one direction at greater or less angles, and do not now show the evidence of any folding in the earth's crust there, other than a close similarity in the appearance of many of the ledges, and an evident recurrence of the same mineral-bearing series. From the latter circumstance it may be inferred that anticlinal folds have occurred, though to make out now, in these partially metamorphosed rocks, any distinct order of stratification or superposition would be impossible. The rocks of Grayson undoubtedly belong to that long period of transition between the unstratified Azoic and that series which, more than any other, has the right to be called the first of fossil-bearing rocks—the Lower Potsdam, or the first of the Cambrian or Lower Silurian. Therefore the mineralogist may not be surprised at finding the following list of minerals, though the existence of some of them is more suspected than positively ascertained as yet :

Garnet, hornblende, kyanite, corundum, and staurotide, all of which are known to exist. Labradorite, magnetite, trap, oligoclase, orthoclase, albite, chloritis, manganesian epidote, rutile, talc, actinolite, mica (phlogopite, biotite, muscovite), quartz, limestone, tourmaline, beryl, asbestos, steatite ; granite of several varieties, including porphyritic granite, syenite, and such metalliferous ores in quantity as copper and iron pyrites, magnetite (as mentioned), specular ore, brown iron ore, and occasionally arsenic, antimony, silver, lead, and gold, with the existence of nickel strongly suspected.

At least two cross-sections would be desirable ; but one will be sufficient to give a very fair idea of the general position of the rocks, the trend being somewhat uniform throughout.

Beginning on the south, about where the Peach Bottom Mountain crosses the Virginia line, we encounter a band of hornblende, slates, steatites, then mica and talc schists, quartzites, etc., which lie between the great Ore Knob Copper Lode and the Peach Bottom Copper Vein; then going north, when in the vicinity of the last-named vein, you encounter talco-mica slates—a broad band—occasionally interstratified with mica schists; then at about half a mile more, over occasional bands of gneiss, with some manganiferous epidote to a broad ledge of soapstone; then over gneissoid strata and a succession of talco-mica slates and schists to the great iron and copper pyrites lode, which makes such extraordinary surface showings in Carroll County. Then for six or eight miles, over a succession of talco-mica slates and schists and manganiferous epidotes and slates in almost any order of succession, with some soapstone, we reach the southern limit of the great granitic bands which show on so large a scale in Buck Mountain, Point Lookout, and that range of rocks. These are about six miles in thickness, then they begin to give way to rocks which are less granitic for the more purely felspathic series, which in turn give way to the hydro-mica slates capping the Huronian system in Iron Mountain; in the last four or five miles crossing that system, which is now yielding gold in interesting quantities farther northeast in Montgomery County.

IRON ORES.

The iron ores of greatest importance in Grayson are the magnetites. *Brown ores* exist toward the southern and southeastern part of the county; but mainly as gossans or hydrated peroxides, resulting from the decomposition of pyrites. The same veins of iron and copper pyrites, which show such immense beds of gossan on the surface in Carroll County, do not present so great a surface showing, in that way, in Grayson. Notwithstanding this fact, there are very

flattering indications of brown ores here and there along the
pyrites vein in the neighborhood of New River, and on the
Southern Copper Lode in the extreme southeastern part of
the county.

Specular ores are found in quite a number of localities in
the county. One vein, somewhat less than six feet thick,
shows in the south slope of Iron Mountain. Its purity has
not yet been determined; but the quantity of the ore must
be very great, as the measure, no doubt, extends for a great
many miles. There are, occasionally, micaceous looking
pieces of specular ore found; but no positive data have, as
yet, been gathered relative to the quantity and the various
localities in which it shows most prominently.

The *magnetic iron ore* of this county is found in veins that
may be said to lie in the northeastern prolongation of the
celebrated beds of Mitchell County, North Carolina. There
are three distinct measures running parallel with each other,
from a locality on the State line a little south of the mouth
of Wilson, pursuing a direction north 75 east through the
Billings land and to the south of Independence, and on to the
eastern limit of the county, crossing New River, the last time,
about three and a half miles north of Old Town. Along the
course of these very valuable veins they hold very different
measures: eleven feet, one of them, at the Billings Mine;
another, to the south of that one, reported nearly one hundred feet thick; again, as you approach the eastern line of
the county, near the river, neither of the three exceeds three
and a half feet in thickness. It is hardly necessary to go
into a more full description of each place at which the ore
of these veins is exposed. Any one can take the map and
readily find the general locality in which they occur. It
might be of interest to describe fully the showing at Billings,
as from it a better idea may be had of the rock material accompanying that vein, which appears here to be the north-

ward vein of this series. The vein at Billings, or the Brush Creek Mines, four miles southwest of Independence, is eleven feet thick, having on one side of it pyrites of iron and copper in valuable proportions (perhaps nickel), roof of hornblende, schists, and slates, floor of same, quartz predominating. In this floor is a seven inch vein of spar, apparently fluor spar. The dip here is 60° southeastwardly, and the trend, which is local at this place, is nearly northeast. The vein matter will yield thirty per cent. of fine ore. Now and then fragments of fine magnetite are found in a line to the north of this series, as at Mason's, of Elk Creek, but no developments have as yet proven the thickness of the vein. The prospect of a fine body of ore being found on the river within three or four miles of Old Town is very good. The character of the fragments obtained there—for instance near Cherry Grove—is of the highest order ; and, judging from the size of the pieces lying on the surface, the veins must be of good dimensions. Two forges in the county have been or are now using ores from these veins : one on Little River southwest from Old Town, and the other, perhaps discontinued now, on New River, eight miles southwest from Independence. Running along with this series of veins, in places, is a seam of *specular ore*, not yet found over eight inches thick.

IRON PYRITES.

It would be interesting to know the number of ledges of rock, besides well-recognized veins and lodes in Grayson, which are more or less heavily impregnated with pyrites. That some of the pyritous-bearing subjects carry gold in good quantity, and now and then nickel, there can be but little doubt.

The strictly iron pyrites, however, is a largely preponderating constituent in three of the great metalliferous measures of the county, namely, the Northern Copper Lode, the Peach Bottom Copper Lode, and the Southern or Ore Knob Lode, which cuts into the extreme southeastern corner of the county.

The Northern (or Iron Lode) is from nine to fifteen feet thick, and shows, as at Hampton's, a pyrotite and copper pyrites combined; in some parts of the vein giving an average of five per cent. of copper, the rest being iron and sulphur having the appearance of holding *nickel*.

The course of this lode is, for a part of its way, through Grayson, northeast. Coming into the county, at a point a mile or two southwest of Dowten's Ford, it continues on by Hampton's close to the mouth of Little River, thence on by a point somewhat more than a mile north of Old Town, in the direction of the great outburst in Carroll County, about fifteen miles length in Grayson. That this sulphureted vein would become an important commercial feature in the county, in case of cheap modes of transportation, there can be no doubt. The openings made at Hampton's and other places amply prove its continuity and size. Along it there is not so much gossan as on the Southern Lode.

The Peach Bottom Lode is charged, for the most part, with copper pyrites only, and its iron pyrites is really too small in amount to require mention.

The Southern Lode seems altogether, in places, to be composed of iron pyrites when you get down low enough below the decomposed ores to strike it. This lode being sometimes forty feet thick, the quantity of ore in it may be imagined.

There are localities in which pyrites is sometimes a large constituent of the rocks—on Elk Creek, Fox Creek, Wilson Creek, and on the streams in the western part of the county in the slopes of the Balsam and White Top Mountains. This is the range of ores which is most likely to yield gold in paying quantities.

MANGANESE.

Manganese is a large constituent in many of the rocks, but is not developed in any distinct veins in sufficient quantities to pay for mining.

LEAD.

Lead is found interjected with the copper ore in the Peach Bottom Lode, and may be valuable on account of the silver it is known to carry.

It is also known to exist in a rock having the appearance of trap, which trends through about two and a half miles south of Elk Creek Post Office.

It is highly probable, if this were followed up, it would lead to something valuable.

COPPER.

This is an unsatisfactory subject to treat, so little is known of the great veins which hold it—at any depth.

Copper pyrites can be detected in a great many of the ledges of rock in the county; but the probability is that it is likely to be found in paying quantities in the Northern Lode passing Hampton's, the Peach Bottom, or Middle Lode, the Southern Lode, and in one of the sides of the Brush Creek or Billings Vein.

In the first or Northern Lode it has already been found to yield, in some places, an average of five per cent. of copper, though this may not hold as a rule. In the Peach Bottom Lode, in this county, no developments of any consequence have, as yet, been made; and it might be premature to assign to it the same character it has at the Peach Bottom Mine in Alleghany County, N. C. On the Southern or Ore Knob Lode in Grayson, it has been inferred, from the live character of the gossan, that there must be a good percentage of copper below.

GOLD.

So important a statement as that there are immense quantities of the precious yellow metal in the county should, no doubt, be made with a great deal of caution ; yet such is the opinion advanced here, after an investigation of some length.

That the gold exists in the parent rock in sufficient quantities to pay for extraction is rather a premature statement to make ; but that the disintegration of these Huronian rocks has, through ages, left paying quantities in the *débris* and drift along the streams, is confidently believed. The region of Elk Creek participates in this distribution. In all likelihood a line parallel with the course of Iron Mountain, about the distance from that mountain that Elk Creek Post Office is, clear through the county, would be in a gold-bearing series. The gold, it may be submitted, results from the decomposition of a pyritous quartz and felspathic band, which seems to follow this line, showing with some distinctness, also, on the bank of New River, a few miles above the old Grayson Sulphur Springs, on the line between Grayson and Carroll.

Elk Creek, however, in the beautiful valley that looks, in spring time, like a jewel from heaven, on the north side of Point Lookout, is the district most likely to pay the intelligent prospector.

The question may well be asked, if this be true, why should so valuable a metal have lain there so long without being found out long ere this ? The same might be said with equal propriety of the Brush Creek gold belt in Montgomery, which the miner, only the other day, never even so much as suspected to contain gold. Now it is attracting universal attention.

SILVER.

Silver has been found by close analysis to exist in the ores of the Peach Bottom Lode ; in what quantities, however, it has not been determined. Also, in the Northern Copper and Iron Pyrites Lode one chemist found it. It is highly probable that deep mining only on these veins will find it in paying quantities.

Then we may also hope that these veins, following the habit of similar ones in Cornwall, will also yield *tin*.

LIMESTONE.

Only one ledge of limestone is known to show in the county. That crosses the turnpike leading from the mouth of Wilson to Marion. It is in sufficient quantities to be highly valuable.

FELSPAR.

There are apparently all the varieties of felspar in this county, the scientific names for which are orthoclase, oligoclase, and albite, comprising the important kinds, such as labradorite, etc. Orthoclase, which is distinctly the potash variety, is quite abundant, sometimes in measures eight feet thick, as at Elk Creek, in several places on the road from Blue Springs Gap to Independence, and at many other points, east and west. From the decomposition of this rock, which contains about 14 pounds of potash to the 100 pounds of rock, and other felspathic varieties, Elk Creek owes the exceeding fertility of its soil.

From these ledges of pure felspar a fertilizer may yet be devised which, if used in connection with the gypsum from the inexhaustible beds of Smyth County, will render it possible, at a very moderate expense, to bring up worn-out or unproductive lands to a high state of fertility.

GRANITE AND SYENITE.

Granite of every variety is found in Point Lookout and Buck Mountains. It may also be found in the spurs of the

Balsam and White Top Mountains, and again far to the east
near the eastern side of the county; but its positions of
greatest purity and compactness seem to be Point Lookout
and Buck Mountains, and their vicinity, even the black and
white being there, the porphyritic, and that which graduates
into the syenitic, finally showing true syenite. As a building
or ornamental stone there are great masses of it which could
not well be excelled.

ASBESTOS.

This mineral exists in some quantity along a belt just south
of the Northern Copper Lode. It has been found in handsome
specimens on Little River and below the Hampton Mine. It
is reported also from Black's, in the western section of the
county, not far from Grant.

SOAPSTONE.

There are great masses of good soapstone following the
range north of Peach Bottom Mountains, extending for many
miles through the county; also in beds still further north.
It will be found useful in furnace lining and for building purposes.

TIMBER.

There is a great variety of fine trees in Grayson. White
pine is abundant in the south spurs of Iron Mountain, Balsam Mountain, and White Top and points along the river.
White oak, chestnut oak, chestnut, etc., are very plentiful.
Timber is so abundant in nearly every part of the county that
some easy means of getting it out would insure a bountiful
supply of very cheap charcoal to the furnace men for years to
come.

WATER POWER.

On account of the invariable flow, at all seasons, of a large
number of fine streams, over rapidly descending beds, this

county must be regarded as having fully as much water power as any in the State. Including the river, which, here and there, has sufficient fall, any force, from 1,000 cubic feet per second down to 10, can be had with but little outlay—Little River, Elk Creek, Wilson Creek, Fox Creek, Peach Bottom Creek, Bridle, the upper ends of Little and Big Helton Creeks, Grassy Creek, Brush Creek, Knob Fork, and two or three others in different parts of the county. It would be idle, as well as unfair, to institute a comparison between, or to attempt a description of, particular locations. All of these creeks afford excellent powers every mile or two for all the purposes of milling and manufacturing, which, taken in connection with the constancy of the streams, renders the county, in this respect, the superior of nearly all the counties in the State.

MANUFACTURES.

There are very few manufactories in the county, except a carding machine or two, and the Little River Forge, now in operation. There are the usual grist and saw-mills in every neighborhood; but, for want of facilities of transportation, there has been no inducement to build any extensive factories of any kind, though the people are sufficiently enterprising.

AGRICULTURE.

The different parts of the county are very diverse in their agricultural features. Wherever the granitic, or more felspathic series predominates, the soil seems stronger and more capable in every respect. Where the streaks of manganiferous epidote and the more easily decomposed talco-mica slates are, there you find an unsatisfactory soil; but happily, as to the great body of the lands, the latter are in the smallest proportion. It is plain to be seen that the soil throughout is the result of the decomposition of the rocks in each locality; wherever the rocks have been chiefly composed of alumina (or clay material), silica, and potash, soda or lime, there are the finest soils. Elk Creek is a good example of this. Nowhere in Virginia can a more beautiful scene be found than Elk Creek Valley affords, looked at from any of its surrounding hills. Its beauty and loveliness are owing to the matchless character of the material composing its soil. With but little attention, here and there, in the steep portions, these lands would never lose their fertility. On the contrary, they ought to increase in productive capacity each succeeding year.

But Elk Creek is not the only beautiful gem in Grayson. The hills, often steep, are crowned with that growth to the summits which indicates the strength of the soil. Frequently upon the tops of ridges nearly 4,000 feet above sea level some of the finest corn-fields are to be seen. Beginning on the Upper Heltons, Grassy Creek, and coming over each succeeding creek and ridge until you reach the eastern limits of the county, there are thousands of acres of land, at from $5 to $15 per acre, which would prove far more productive than twice the same quantity in localities farther east. There is nothing raised in this latitude which these lands are not capable of producing in abundance.

SCENERY.

A valuable collection of fine views could be obtained in this county. Whether landscape scenery, such as that presented by the incomparable scope along Elk Creek, or extended views from high mountains, or the fine pictures made up by waterfalls over 100 feet high in mountain dells strewn with massive rocks, all the same could the lover of the beautiful in nature find every sense gratified. The river, as it winds back and forth between the high hills, offers many lovely views. There are few localities in the county that cannot produce some well-known point of interest in the line of scenery. Many of them if painted and exhibited would command the admiration of the most experienced and critical observers.

FRUITS.

The apples of Grayson have commanded for years a very wide fame in the surrounding country for their flavor and excellence.

While the apple seems to be in its *native* home here, the peach, quince, pear, cherry, and plum are regarded as sure to make a crop each year in Grayson as in any other locality.

Grape culture could be carried to high perfection in the county, judging from the abundance of native varieties. There has been but little effort made to improve the old or to introduce new varieties. Diseases of any kind seem never to have attacked the grape in Grayson. No doubt the soil, and the large area of southern exposures, would make grape culture very successful if native varieties were used.

Bee culture could be brought to a high state of perfection on account of the large number of flowering plants and trees, and the abundance of moist places for the bees to use in the hollows.

Fish culture, after the county is rendered more accessible by railways, will become an important feature, not only for mere sporting purposes, but from an economic point of view. The streams are peculiarly well adapted to every species of game fish. The mountain trout is now very common in nearly all the streams, and in some of them affords excellent sport. The New River catfish reaches its highest perfection in this county. Unlike his namesake in the western waters, he is here regarded fully equal to the best table fish in excellency of flavor and all other good points.

TRADE IN CATTLE, SHEEP, WHEAT, CORN, AND TOBACCO.

The possible trade of this county in all the staples would be difficult to approximate under the favorable conditions of

cheap and abundant transportation. Besides being naturally
a good cattle and farming county, Grayson is capable of
making good tobacco. The record of the number of cattle
and sheep now annually sold is no just estimate of her capacity; for, with the stimulus resulting from increased means of
transportation, without clearing any more land even, the
county would increase its revenues from all classes of produce fully tenfold, if not more.

There has been great improvement of late years in cattle
and sheep. Elk Creek and Bridle Creek seem to have led
off in this direction. Shorthorns in cattle, and Cotswold
and Shropshire-downs in sheep, seem to be the favorites.

The county sends about 1,100 cattle every year, which,
directly or indirectly, make their way to European markets,
and her total sale of

Cattle annually is about. . . . . .3,800 head.
Sheep   "   " 4,000   "
Wool, mostly used up at home, about .9,000 pounds.

The quantity of wheat, corn, and tobacco now being exported is scarcely worthy of notice.

Grayson sells annually a large quantity of bacon; but is
now doing but little with horses, except improving the stock
in some quarters.

TOWNS AND VILLAGES.

The county site is Independence, situated somewhat east
of the center of the county. It has the usual number of
hotels, churches, stores, saddleries, smith-shops, etc. The
Grayson *Clipper*, a progressive weekly newspaper, is published there. Old Town, once the county site before the
county was divided, is situated toward the southeastern end
of the county, near the Carroll County line. It is about one
mile south of New River, and is the center of a good trading

and mining region. There are stores, churches, a hotel and post-office there. Mouth of Wilson, Elk Creek, Bridle Creek, Grant, Greer's, Carsonville, and one or two other points are good trading posts, and now annually collect and send off, besides other produce, a large tonnage in medicinal herbs, roots, etc.

The *public schools* are being better protected by the State government, and are gaining greatly in the estimation of the people.

ASHE COUNTY, N. C.

It would ill become any one to attempt to give a thorough *résumé* of the mineral resources of any part of North Carolina, after that work had once been done by such men as KERR, HUNT, and GENTH; but, as the counties of Ashe, Alleghany, and Wautauga are directly in a continuation of the great and massive belts which pass through Floyd, Carroll, and Grayson in Virginia, a feeble attempt to show this continuation may well be excused.

Ashe County seems to exemplify all of the best that may be said with respect to the series of rocks of which it is composed, besides presenting a rich and charming scenery of unsurpassed loveliness in its lofty lone mountains and romantic gorges. Lying as it does at the head of nearly all the great rivers that flow in every direction from it, its elevation gives it a summer climate unequaled for health, to which crystal freestone water from thousands of never-failing springs lends a security far beyond the conception of any one who has never felt its influence.

Ashe lies in the plateau of the Blue Ridge and Unaka Ranges, having one on the south and southeast boundary, and the other on the northern and western sides, having Grayson, Va., on the north, Alleghany, N. C., on the east,

Wilkes and Wautauga, N. C., on the south, and Johnson County, Tenn., on the west.

HOW WATERED.

The county is excellently well watered by the North and South Forks of New River, which yield a water power for every mile of their course of great reliability, and of any desirable volume.

GEOLOGICAL.

To use distinctions which are more or less arbitrary, Ashe shows the rocks of the Laurentian and Huronian epochs, sometimes placed under the general appellation of the metamorphic series.

They include granite, gneiss, syenite, quartz, hornblende, mica and talc, chlorite, mica and talc schists and slates, gray soapstone, epidote, felspars of all varieties, trap, zeolites, actinolite, and many others, the lithology of which would be entirely obscure to the general reader. Pure granite is rare in the county; but, toward the southern side principally, there are broad bands of gneiss, as those showing about Ore Knob. As you go north of this there are extraordinary bands of hornblende, followed by mica and talc slates, soapstone, epidote, etc., in recurring series, all pursuing a course nearly north 70° east and south 70° west, with a dip varying between 30° southwardly on the north side to the perpendicular toward the south side of the county. In the gneissoid system, nearer the Blue Ridge, are the copper veins in which Ore Knob and Copper Knob are situated; while, following the hornblendic series near the northern middle part of the county, are the copper veins in which Elk Knob Mine and Phœnix copper ores are situated. Toward the northern side, on and near to the North Fork of New River, are the great bands of magnetic ore in the epidotic series in part, and in the gneissic and hornblendic running along with it and just south of it.

COPPER.

The copper range in Ashe is admittedly one of the best and most reliable in the world. The Ore Knob Mine has sufficiently demonstrated this point, the quantity of cement copper and ingot produced since 1872 not being far from 25,000,000 pounds; last year having shipped 2,436,392 pounds of ingot copper. The energetic and efficient management there having erected, in a surprisingly short space of time, a singularly efficient and entirely satisfactory plant, including machinery and furnaces, the capable mine soon began and continued to yield returns that have fallen at only rare intervals below the expectations resulting from the thoroughly practical tests of the management, sustained by the no less excellent investigations of Dr. HUNT.

This mine lies in what is familiarly known in Virginia as the Southern Lode, its continuation in Virginia showing excellent copper ore at the Toncray Mine in Floyd County, Va., and immense quantities of gossan near Sparta. Again, to the southwest of Ore Knob, at one or two points, as Mulatto Mountain, this lode shows strong surface indications. Some persons are rather inclined to the belief that Copper Knob, or Gap Creek Mine, is also in this lode ; but a careful examination proves it to lie south of the Ore Knob Lode.

This vein or lode, as it shows at Ore Knob, is declared by KERR to be the most remarkable of the many copper veins showing in North Carolina. He says :

That though it was opened before the war, it was not until it fell into the hands of the present owners, in 1871, that it began to show its real character. "These gentlemen have opened the vein by a series of shafts and tunnels, and have been repaid by the discovery of a body of ore which is not equaled by any mine I know of outside of Ducktown. . . .

14

"The rock of the region is a gray and usually thin bedded gneiss, with mica schists and slates. These have a prevalent strike a little east of northeast, and dip east at a tolerably high angle; though both dip and strike are subject to considerable variation. The walls of the copper vein are micaceous gneiss and mica slates, with a strike north 57 east, and dipping southeast at an angle 40 or 45. The copper vein is coincident in strike with the rocks, but is vertical in dip, cutting across the strata, so that it is a true fissure vein, and not bedded like those at Ducktown. It is traceable by an outcrop of gossan for more than a mile, and has been proved by trial shafts and trenches for nearly 2,000 feet. The breadth of the lode varies from 6 to 15 feet (is stated to measure 20 in some cases, *which is too*), averaging about 10 probably." PROF. KERR then goes on to speak of the number of shafts then sunk at the time of his visit, etc. Now there are eight shafts over a length of about 800 feet, the principal of which are the Engine Shaft on the crest of the hill, and Nos. 2 and 3 south. The mine has been carried to a depth of 350 feet. PROF. KERR goes on to say : "There is, properly speaking, no gangue stone, the whole breadth of the fissure being filled with ore. The gossan, which is decomposed oxidized ore, extends to an average depth of over 50 feet in the different shafts, the lower half containing, however, a valuable percentage of copper in the form of oxide and malachite. Below this level of oxidation the ore is sulphuret of copper." DR. HUNT gave the gossan yield at 14 to 22 per cent. of copper, and the sulphurets of iron and copper were last winter yielding an ore which assayed fully as much. An inspection of the sketch and section on the adjacent pages may perhaps lead to a better understanding of the immense amount of work done at this mine. The nearest furnace showing in the sketch is a reducing furnace, the next one (near the end of the railway) is a reducing furnace, and

also holds the refining furnace. The large building at the
upper end of the railway is the ore-house, holding the crush-
ing machinery, etc., while beyond it is the sky-house, which
is erected over the engine shaft.

The operations about this mine for the last nine years have
had the effect of creating an immense business throughout
that region on a paying basis. The amount of money annually put in circulation for labor and supplies must be very
great.

COPPER, AND GOLD, AND SILVER.

The ores taken in Mulatto Mountain, from what has been
commonly accepted as the southwestern continuation of the
Ore Knob Lode, yield an average as follows: Copper, 3 per
cent.; gold, $2.05 per ton; silver, $2.80 per ton. These ores
were taken from a vein 5½ feet thick; but it is barely possible
that this vein lies about a half mile to the north of a point
where the Ore Knob Vein should be, if continued. Mulatto
Mountain is ten miles southwest from Ore Knob.

Copper Knob Mine, situated in the Blue Ridge, near the
Ashe-Wautauga line, was found, upon a close examination, to
agree very fully with KERR's description, which is as follows:

"This is a quartz vein. or rather a group of them; the
principal one carrying variegated copper, with a little chalco-
pyrite, malachite, chrysocalla, specular iron pyrite, together
with visible free gold and silver. The vein is in a large body
of hornblende slate, though the prevalent rock of the section
is a gray gneiss, with a strike north 60° east, and dip southeast 40°. The vein is a true fissure, with a direction north
35° west; dip, northeast 45°. DR. EMMONS, who visited the mine
when it was open, says, 'This is a true vein, and has a perfect
regularity in direction, as well as in its walls.' The width
is variable, being 18 inches at the surface, and from 12 to 24
inches at different depths below ground." The ore, analyzed

by Mr. Manross, gave "gold, 1½ ounces, and silver, 18 ounces per ton of mixed rock and ore." Handsome specimens of purple copper ore from the center of the vein, showing much free gold to the eye, yielded about $2600 in gold to the ton.

The next great series of copper deposits in Ashe lies four miles north of, and nearly parallel with, the Ore Knob Lode, about the middle of the county. This line of ores is in the southwestern continuation of the Peach Bottom Lode; chiefly shows in a decomposing micaceous gneiss, and toward Elk Knob in hornblendic strata.

Near Jefferson, at Weaver's, and at Foster's, near Phœnix Mountain, there are fissure veins cutting the strata. They are accompanied with quartz, and often show a thickness of nine feet. The average yield of these veins is very hard to determine from present workings, but much of the mass is a very pure sulphuret of copper. Occasionally, copper glance is found.

Elk Knob Copper Vein, really in Wautauga County, seems to be nearly at a point where the Peach Bottom continued that way would strike, but it is in an entirely different kind of rock from Peach Bottom Vein. Elk Knob Vein has been very well exposed by the owners in several places, and is at different points variable in thickness; in one deep ravine being seven feet, and, at another point, sixteen feet thick. The vein seems to be largely composed of mundic; but yields fine specimens of gray copper ore and copper pyrites, all mixed with a low percentage of gold and silver.

This vein is undoubtedly a valuable copper vein, but has been badly prospected. If due regard had been observed as to the carbonate of copper showing in the gossan at different points, and the shafts sunk accordingly, much more satisfactory results would have been achieved. As it is, the work was done wherever the vein was most accessible, and, unfortunately, the poorer alternations in the vein were thus exposed. There are extraordinary surface indications along this vein for some miles in length—mostly gossan or oxidized ore.

Then, again, on the north side of Phœnix Mountain, there are evidences of the southwestern continuation of what is known in Virginia as the Northern, or Hampton Copper Lode. No developments of consequence have ever been made on it in Ashe; although, now and then, gossan may be detected along it in considerable quantities.

There are other places in the county where copper ores have been found, sometimes in flattering quantities, as at Wither-spoons, the Old Meat Camp Mine, etc.; but the thickness of the veins has generally not been sufficient to warrant much outlay.

IRON ORE.

The iron ore in Ashe of greatest value is found in the magnetic bands on, and north of, the North Fork of New River. At Ballou's, on North Fork, the great magnetic vein was found to be massive ore, accompanied with hornblende, talc, mica, and, occasionally, tremolite-trap and quartz as insoluble constituents, in the proportion of nearly 20 per cent. of the vein, at a point where the vein is 30 feet thick; trend, north 55 east. Just south of this vein, about 250 yards, is another vein of the same ore, separated from the first vein by a material which is mainly hornblende schist, mica slate, and schist and gneiss, general trend being south 55 west. This vein is 5 feet thick; while the larger, though for 300 feet 30 feet thick, is generally not over 15 feet—sometimes 12. These veins are continuous, either way, for many miles. These ores contain about 0.026 of phosphorus, and no titanium, by John Fulton's analysis.

Then, again, in the gneiss, etc., of Helton and Horse Creeks are massive ores, coarse, granular, and highly magnetic.

PROFESSOR KERR says, on Helton Creek, six or eight miles
east of the Horse Creek ores, "are still larger deposits of very
pure magnetic ore which has been long used in the forges of
the neighborhood. The ore is a close-grained and very pure
magnetite, one of the beds of which is reported to be eighteen
feet in thickness, and another nine feet."

Toward the junction of the North and South Forks of New
River, in the northeastern prolongation of the Ballou veins,
a forge has been running for some time, making a bar-iron of
the highest quality.

Limonites, of great purity, are common in the gossans of the
different copper veins. Ore Knob, Elk Knob, and the north
side of Phœnix Mountain show very considerable deposits.
The south flanks of the Balsam and White Top Mountains
show specular ores, but the quantity is not easily ascertain-
able.

MICA.

Large-sized mica is found abundantly in Ashe, in a line of
dikes composed of felspar, quartz, and mica, pursuing a
course through the central part of the county, northeast and
southwest, at an angle with the strike of the rocks. The
largest developments yet made are at the Little Mine on the
South Fork of New River, at Harden's, and at places on the
head of Three Top Creek. There are also large pieces of
mica reported from the south spurs of Balsam Mountain and
White Top. The two veins at the Little Mine are respectively 30 and 18 feet thick, and are apparent for about one eighth
of a mile, though there is no doubt of their continuation for
miles either way.

KAOLIN.

Fine kaolin, resulting from the decomposition of albite in
these mica veins, is very plentiful.

Felspar.—Very pure felspars exist in large quantities in the dikes holding large mica.

Asbestos is found in a line of rocks three miles north of Jefferson, the county site, but the quantity is not great. In the south spurs of White Top Mountain, near Black's, it is said to be in quantity and of good quality.

TALC.

A nearly pure talc or steatite is found in many sections of the county, but that found in the east face of Elk Ridge is preferred at Ore Knob Copper Mine for furnace lining. It is easily sawn into blocks of any desirable size.

TIMBER AND CHARCOAL.

The timber of the county is of every conceivable variety known to the latitude, and in forests of unlimited extent. Charcoal, except in the immediate neighborhood of the copper furnaces, could be had in great abundance for years to come at a merely nominal figure.

The lines of transportation through Ashe, as to railways, are only as yet projected. Possibly the Statesville and Virginia Railroad may, at some early day, be built, passing near Ore Knob; and there may be others of which the writer knows nothing.

ALLEGHANY COUNTY, N. C.

This county is much like Ashe in its geological features; lying just east of Ashe, it holds the northeastern continuation of its copper veins.

The Peach Bottom Copper Lode shows to best advantage at Peach Bottom Copper Mine on Elk Creek in this county. It now shows about nine feet of ore in walls of a highly micaceous gneiss, or mica slate, sometimes talcose. The ore is usually copper pyrites; occasionally purple copper ore, with a considerable admixture of galenite. It is claimed for this mine that it will yield largely, also, in nickel, antimony, and arsenic. This vein has a dip southwardly approaching the perpendicular, and a strike east of northeast. It is continuous for miles. Has been actively developed to a depth of over 150 feet within the last year.

On the Ore Knob, or Southern Lode, south of Peach Bottom Mountain, are the great deposits of gossan or limonite, hundreds of feet in extent. They are close to Sparta, the county site.

Nearer the northeastern border of the county are valuable deposits of *asbestos*.

THE COUNTIES OF SOUTHWESTERN VIRGINIA.

Augusta County, 1738, from Orange County. Comprising all west of the county of Frederick and west of the Blue Ridge. This territory, of which nearly all of the counties of Southwest Virginia are now composed, and our southwestern counties, made a part of the territory alluded to by Gen. Washington, when he spoke of "the mountains of West Augusta."

Fincastle was formed in 1772 from Bottetourt, and was extinguished in 1776 by the formation of Washington, Montgomery, and Kentucky counties.

Montgomery, 1776, from Fincastle County.

Pulaski, named after Count Pulaski, was formed in 1839, from Montgomery and Wythe counties.

Wythe, 1790, from Montgomery.

Grayson, 1793, from Wythe.

Washington, 1776, from Fincastle County.

Russell, 1786, from Washington County.

Lee, 1792, from Russell County.
Tazewell, 1799, from Russell and Wythe.
Giles, 1806, from Monroe and Tazewell.
Smyth, 1831, from Washington and Wythe.
Floyd, 1831, from Montgomery.
Carroll, named after Charles Carroll of Carrollton, 1842, from eastern part of Grayson.
Wise, 1855, from Lee, Scott, and Russell.
Buchanan, 1858, from Tazewell and Russell.
Bland, 1861, from Wythe, Tazewell, and Giles.
Dickenson, 1880, from Buchanan and Wise.

CENSUS ITEMS.

Population of the counties by the census of 1880 compared with that of 1870, from information kindly supplied by Gen. Walker, Superintendent of Census:

	1880.	1870.
Montgomery	16,693	12,556
Pulaski	8,750	6,538
Wythe	14,318	11,611
Smyth	12,159	8,898
Washington	25,203	16,816
Giles	8,794	5,900
Bland	5,004	4,100
Tazewell	12,861	10,791
Russell	13,906	11,103
Scott	17,233	13,036
Lee	15,116	14,100
Wise	7,772	4,785
Buchanan	5,694	3,775
Floyd	13,255	12,000
Carroll	13,323	9,147
Grayson	13,068	9,597

14*

www.ingramcontent.com/pod-product-compliance
Lightning Source LLC
Chambersburg PA
CBHW021403210326
41599CB00011B/987